Crude Oil Waxes, Emulsions, and Asphaltenes

Crude Oil Waxes, Emulsions, and Asphaltenes

J. R. Becker

PennWell Books
Tulsa, Oklahoma

Copyright © 1997 by
PennWell Publishing Company
1421 South Sheridan/P.O. Box 1260
Tulsa, Oklahoma 74101

Becker, J.R.
    Crude oil waxes, emulsions, and asphaltenes/J.R. Becker. p.   cm.
    Includes index.
    ISBN 0-87814-737-3
    1. Petroleum—Refining—Dewaxing. 2.Emulsions. 3. Asphaltene.
I. Title.
TP690.45.B43 1997
665.5'3—dc21   97-31816
                                                                         CIP

Printed in the United States of America

1 2 3 4 5 99 98 97

Contents

Section II Waxes   101

Section III Asphaltenes   209

Section I

Crude Oil Emulsions

Introduction

Emulsions are among the many problems encountered in the production, transport, and refining of crude oil. Dealing with these complex structural arrangements accounts for much of the expense incurred by oil companies in their daily operations. The presence of water in oil (and oil in water) costs the producer, transporter, and refiner in several ways. When water is present in produced oil several other costly byproducts of its presence result.

Corrosion, scale, and dissolved metals are three important byproducts of the presence of emulsions in produced crude oil. Each of these individual problems must be addressed by the producers prior to the transport and refining of the crude. With increasing environmental regulations, the requirements for safe disposal of the produced water derived from the resolution of these emulsions are also increased. Thus, the cost of resolving these problems escalates, and the need for understanding their nature becomes critical to the operations of an oil company.

The information presented here is not intended to be an exhaustive discussion of the subject of emulsions, but rather a discussion directed to the particular aspects of these systems that relate to the oil industry. There is a fair amount of chemistry, physics, and mathematics involved in this subject, but efforts have been made to minimize the use of rigorous treatments of these areas. Throughout this book the approach is to develop an intuitive discussion that has practical meaning to those faced with the resolution of these problems.

1
Petroleum Companies and Emulsions

The average consumer of petrochemicals is blissfully unaware of the heroic efforts required to provide the object of his requirements. Although he may realize that providing gasoline for use in his car is a little more complicated than parking next to an oil well, hooking up, and filling up, he cannot imagine its being much more complicated. Drilling and work-over rigs, offshore platforms, pipelines, pumping stations, treating plants, and storage facilities are eyesores to the uninformed, rather than the marvels of engineering they really represent. Most consumers know that there are such places as refineries, but these complex monsters loom in the backgrounds of their communities as curiosities. The enormous complexity of these structures and the services they provide remain a mystery to the uninformed.

Oil and Water

The process of lifting crude oil from a reservoir is a complicated procedure that requires a substantial amount of engineering. More often than not, it is considerably more complicated than the simple process of drilling a hole. Considerations of several variables must be incorporated in the lifting methods employed by the producer. The following list represents just a few of these variables: the type of geologic strata, reservoir pressure, gas and/or water drives, crude oil viscosity (API gravity), fines migration, water content

and composition, scaling tendencies, corrosion potential, sour gas (H₂S) concentration, asphaltene and paraffin wax concentration, and emulsion content.

Several of these considerations involve the presence of water (e.g., drive, scale, corrosion, and emulsion). Whether it is emulsified or free, its presence dramatically impacts the efficiency of production realized from the reservoir. Although free water is very common in crude oil production, its presence is of a lessor concern to the producer than the emulsions of water in oil, or oil in water. Free water is easily removed by settling procedures, but emulsions often require the addition of heat and/or chemicals to separate them from intimate contact with the crude oil.

Emulsions

Emulsions are a significant concern to the producer, transporter, and refiner of crude oils, since their inclusion in the product of each reduces the overall profit they may realize. To the producer, the cost of treating emulsions to reduce the water content of the oil represents a significant investment in energy, equipment, and chemicals. Ideally, the water included as emulsions in the crude oil should be resolved as early in the production process as possible (e.g., down-hole), but this requires the ability to introduce the chemical into the well.

This requirement has a significant impact on the completion costs of the well, since the chemical must be injected down-hole. Thus, provisions for chemical injection must be included in the equipment design of the producing well (e.g., the placement of capillary tubing alongside of the well tubing). Expensive chemical injection pumps capable of bucking well pressures must also be allocated to production string design.

Chemical storage facilities, heater treaters, water draw-down facilities, and down-stream settling tanks must also be incorporated into field production design so that water can be separated from the crude oil. Because the large majority of production wells in operation before the 1970s were engineered without the provision of chemical injection strings, considerable costs would be involved in their placement. Thus the costs of reengineering these facilities are very seldom justified (see Fig. 1–1).

Production Environments

Environmental conditions are of paramount concern to the producer, and average ambient weather conditions very often determine much of the design specifications of the production facility. Oil production locations are often found in some of the harshest conditions of climate on earth, and as such require innovative solutions to problems very few industries ever face. Extreme cold presents a host of problems to the producer, problems that must be addressed before successful production can be achieved.

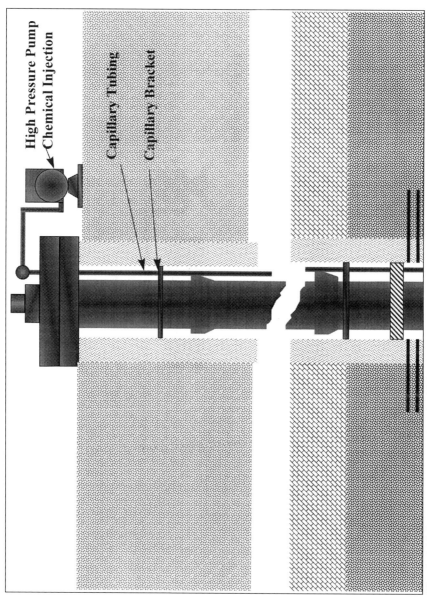

Fig. 1–1 High-pressure-pump chemical injection

There are several crude oil production zones located within the Arctic circle from Alaska to Siberia, and the type of produced fluids derived from these sites represents a wide range of physical behavior. High-wax crudes are common in Siberia, while low-wax oils are more common to Alaska; however, both areas are faced with emulsion problems. On the North Slope of Alaska the temperatures drop to -60° F in the winter, but provisions have been made to deal with these harsh conditions. Enclosed treatment plants and chemical storage facilities have been constructed to provide shelter from the harsh environmental conditions. The expense of providing shelter from these harsh conditions is justified by the high production volume of these zones.

However, there are several producing regions in the lower-48 states that experience similar conditions of extreme cold and the provision of shelter is out of the question. These areas are spread over thousands of square miles and produce only fractional amounts of oil compared to the North Slope of Alaska.

Specialty Chemicals and the Environment

It is not difficult to imagine the complications that arise from the lack of shelter when taking into consideration the physical properties of the chemicals needed to treat emulsions, scale, corrosion, wax, asphaltene, and hydrogen sulfide (H_2S). The physical placement of the chemicals used to treat problems encountered in the field generally requires that the chemical remain liquid so that it can be pumped. This requirement places demands on the formulation of chemicals that are often difficult and sometimes impossible to achieve. Producers want the most cost-effective solution to their field problems, and the added difficulties faced by the specialty chemical companies in their provision is of little concern.

Formulating a chemical blend to withstand temperature of -60° F, without freezing, is a difficult process, especially if the active ingredient freezes at greater 150° F. Specialty chemical companies are commonly faced with providing chemicals that are dramatically reduced in active components in order to fulfill the freezing point requirements for field use. The additional cost of exotic solvents and transportation of dilute products over long distances become considerable factors in the prices specialty chemical companies must charge the oil producer.

Oil-in-Water Emulsions and Environmental Concerns

Once the oil is produced from the formation, the producer must address a host of additional problems. Some of the most demanding problems involve the restrictions placed on the producer by various governmental agencies that regulate the discharge of oil into the environment. These restrictions add significant costs to the production of the crude oil and have, in many cases, led to the closure of production areas and the loss of large numbers of jobs.

Although the cost of treatment significantly impacts the ultimate profitability of a given area, all reasonable means of oil removal are employed by the producer. Most free water coproduced with the crude oil contains significant amounts of oil in the form of oil-in-water emulsions, and when water-in-oil emulsions are chemically resolved, the water phase often contains significant quantities of oil.

The removal of the oil from the water phase often requires the utilization of specialized equipment and chemicals. Additional tanks and chemicals are required to facilitate the separation of the oil from the water, and in some instances specialized flotation cells are employed to meet the regulated specifications of purity prior to disposal. It is ironic that the quality of the effluent water is, in many instances, held to a higher standard of purity than the water into which it is delivered. A balance between increased revenue realized by the retrieval of additional oil from produced water versus the cost of treatment seldom favors production profitability. The reverse situation exists with regard to water-in-oil emulsions, as we shall see in the next section.

Water-in-Oil Emulsions and Economic Concerns

The economic reasons for the resolution of water-in-oil emulsions are very well justified, and represent significant cost benefit to the producer. The relationship between the producer and the transporter is a critical factor in determining the profitability of a production area. While both production and transportation may be separate units of an integrated company, each must meet the requirements of the other to maximize the overall profitability.

The sale of wet oil (emulsion containing oil) from the producer to the transporter is rigorously controlled. The reasons for this control involve the cost to the transporter of the additional potential for corrosion and scale formation throughout the transport network. Purchase prices paid to the producer by the transporter are negatively impacted by the presence of water, and rejection of wet oil is an all-too-common occurrence. The transporter faces similar restrictions when selling his product to the refinery, and the profits realized are also negatively impacted by the presence of water.

Field Application of Emulsion Breakers

A critical variable in the effectiveness an emulsion breaker is the contact time it has with the emulsion. Therefore, the introduction of the emulsion breaker into the system should be as early in the production phase as is possible. As mentioned above, the ideal chemical introduction should be

down-hole, but failing this option, placement at the wellhead is the next best location. This placement option requires that a chemical pump and chemical injection line be placed in the well transfer line. The power and size of the chemical injection pump is then determined by the wellhead pressure it must buck to inject the chemical.

Because production sites are not always equipped with electricity, optional means of powering the injection pump must be employed. In areas where diesel motors power pump jacks, cam or crank assemblies are often employed. These devices consist of a small fly wheel with a radially placed rod that drives the pump, and they are typically placed on the pump jack beam. In areas where production is driven by natural or man-made gas or fluid, alternate means of chemical placement must be devised. These alternate means are generally not as effective as continuous injection, and require the periodic placement of chemical by treatment trucks (see Fig. 1–2).

Individual wells often exhibit individual emulsion problems, since the conditions of production are variable from well to well. Therefore, emulsion problems existing in individual wells are most properly addressed by tailoring the chemical treatment to each well. This may not always be practical, however, and placement of chemical treatment programs at gathering facilities might make better economic sense. This practice places additional demands on the chemical treatment program, and may result in a trade-off between cost and effectiveness.

Although uncommon, secondary chemical addition will sometimes be required at these gathering facilities even though chemical was added at the individual well sites. The physical actions of shear, cooling, and pressure drops experienced by the transported fluids may infrequently cause new emulsion formation. Once the various crude streams are comingled, they are placed in settling tanks, where the water is allowed to separate from the oil.

Production volume and storage constraints add to the demands placed on treating chemicals, since contact exposure time is critical to the effectiveness of the treatments. High volume production sites, like those found offshore, represent special cases. Since storage facilities and special treating devices take up valuable space, they must also provide the highest treatment efficiency. Chemicals must, therefore, assist these treatment facilities in the separation of water from oil, and oil from water.

Specialized Equipment

The mechanical techniques for the removal of water from oil by the resolution of water-in-oil emulsions have not seen many improvements in equipment design over the last several years. Production field designs generally incorporate settling tanks or ponds, and possibly heater treaters when low gravity crude oils are produced. The crude is delivered from the header system to the tank farm, where the resolving emulsion is allowed to separate, and the water is drawn off for further treatment.

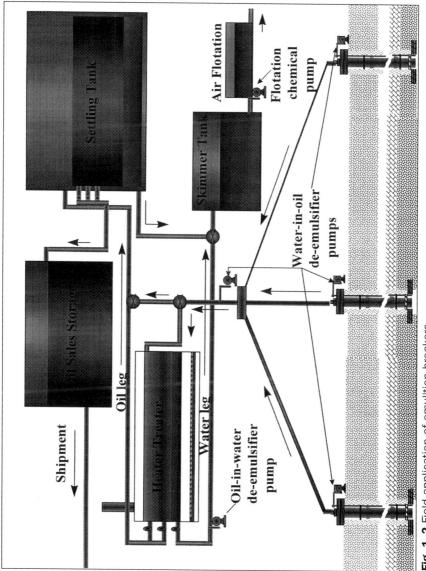

Fig. 1–2 Field application of emulition breakers

If the comingled fluids maintain temperatures that support the de-emulsification process, simple skimmer tanks or ponds are adequate. However, if the comingled fluids are at a low temperature, or the API gravity is low, heat treatment may be required. These heater treaters are often designed to utilize the gases removed from the produced fluids by the gas plant. The treated fluids are then heated to aid in the water separation process, and the dehydrated crude is delivered into a storage tank for sale, while the separated water is sent to additional treatment facilities. It is the water leg of the production stream that requires the majority of the specialized treatment equipment, and it is this sector that represents the least cost benefit to the producer.

The equipment used for the resolution of oil-in-water emulsions has, for many years, centered around the quiescent skimmer tanks or ponds. Again the resolution of the emulsion is facilitated by the longest possible contact exposure time. The water leg, which comes from the dehydration unit, is injected with a reverse de-emulsifier (oil in water de-emulsifier) and delivered to the quiescent skimmer tank or pond, where the oil is skimmed from the surface. The skimmed oil is added back to the head of the dehydration stream and treated along with the incoming comingled fluids.

The oil in water emulsion resolution process very often requires further treatment to remove additional amounts of oil. This secondary process usually involves the combined effect of a flotation aid (high-molecular-weight polymer) and a flotation cell. Although there are several designs available in flotation cells, the generally accepted methods involve the introduction of finely divided air bubbles into the fluid, which aids in the rise of the chemically flocculated oils (see Fig. 1–3).

This treatment equipment usually includes a skimming method for the removal of flocculated oils. The incorporation of this type of equipment into field operations is constrained by several critical variables, among which production volume is of prime importance. If the flotation unit is to be effective, it must be of sufficient size and efficiency to allow maximal production operations; otherwise, it will have a negative impact on the operations profitability.

The Cost of Wet Oil

When the de-emulsification process is unsuccessful, the costs to the producer are not a simple matter of lost crude sales. Remedial processes, with their attendant costs, are also part of the economic picture. Wet oil means that the transporter will not take delivery of the crude until his specifications are met. It also means if remedial methods are unsuccessful, the producer may have to pay to haul it off. These are not trivial concerns either for the producer or the specialty chemical company responsible for the crude oil's de-emulsification.

Battery upsets are common enough that a host of ancillary service companies exist to handle just such problems. Vacuum and hot oil trucks travel

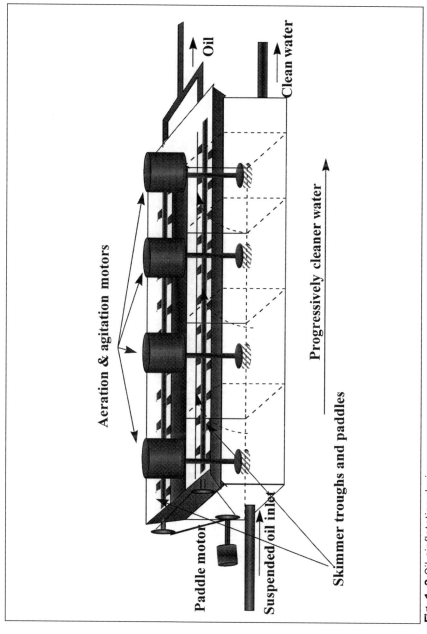

Fig. 1–3 Oil air-flotation device

from field to field in response to the needs of the producer. Vacuum trucks are frequently called to remove wet oil from production areas, while hot oil trucks are used to circulate unresolved emulsions throughout storage facilities to provide additional heat for the de-emulsification process. Meanwhile, the specialty chemical company scrambles to find the source of the problem, solve it, and put the chemical treatment of the field back in order.

The Cost of Oily Water

Although the cost of wet oil is substantial to the producer, the cost of oily water discharged into the environment is high for everyone. The impact of the discharge of oil into the waterways of the world has an effect on the total environment. No agency of any government, or any industry throughout the world, is more sensitive to this fact than the oil industry. Herein lies one of the conundrums faced by the petrochemical industry. Oil producers, transporters, and refineries are businesses, and as such, have a responsibility to the stockholders to show a profit. When the cost of protecting the environment added to the normal operational costs of providing petroleum products to the consumer exceeds a certain limit, the operation must be discontinued. To do otherwise would be flirting with economic or environmental disaster.

The pressures for profitability have led some foreign producers to either circumvent or ignore entirely some environmental regulations. Raw crude or heavily oil-contaminated fluids are discharged overboard from platforms or gathering facilities. Flotation cells and separators are routinely bypassed, and the results are devastating to the environment. As mentioned above, certain governmental agencies in the developed countries have placed such extreme restrictions on production facilities that no choice remains but to close-in operations. To do otherwise would be flirting with economic or legal disaster, because the fines would add to the cost of continued production.

Transport and Refining

The transport of crude oil from production areas to refineries requires a complicated network of high-capacity trunk lines fed by numerous tributaries. The cost of these major arterial lines is substantial, as is the cost of their maintenance. Protection from corrosion is a major concern to the transporter, and as a consequence the controls regulating crude oil water content are very stringent. Therefore, pipeline companies demand strict compliance to a maximum allowable water content. Additionally, the pipeline companies are monitored by the refineries, which also maintain controls on the amount they will pay for variable quality crudes.

Although refineries also accept lower grade, high-water crudes from vacuum truck companies and other sources, they also pay a substantially reduced price per barrel, since special storage facilities are required. Refinery tank farms segregate crudes based on several variables, among which are API gravity, sweet (low H_2S), sour (high H_2S), water content, and scaling tendency.

As mentioned above, emulsion containing crude oils will often exhibit at least one, if not several, of these undesirable traits. Crude oils containing unresolved emulsions frequently produce sludges upon prolonged periods of standing, and these sludges result in additional costs to the storage operations of a refinery. Storage tanks with a 100,000 Bbl-capacity (or more) are not uncommon in refinery tank farms, and the presence of from 4 to 5% sludge represents a significant problem. Refineries are monitored by various governmental and environmental agencies, and in some developed countries, they receive strict fines for the release of potential ground water contaminants. Although the majority of storage facilities are completely enclosed metal tanks, under-corrosion resulting from high water concentrations is an all-too-common occurrence. Thus, the potential for release of contaminants is a real concern to the refinery operations.

Occasionally, refiners must empty these massive storage tanks and perform a program of sludge removal, inspection, and repair. These clean-up, inspection, and repair procedures often require several months and cost millions of dollars. Specialized tank-cleaning companies must be called in to remove the sludge, recover some of the oil, and provide for the disposal of the contaminants. De-emulsifiers play a key role in the clean-up process, and are used in substantial quantities by these specialized tank cleaning companies. So here again, the need for efficient emulsion resolution impacts the effectiveness of tank-cleaning operations.

Refinery Influent Streams

Refinery tank farm operations are directed by the needs of the refinery production requirements, and the percentage of emulsion containing low-quality feed stocks incorporated into the gross feed are strictly controlled. After the lighter gases have been segregated from the feed stock, the liquid fractions are sent to the dehydration and desalting facilities for treatment prior to the crude oil distillation processes. De-emulsifier chemicals are again introduced at the inlet side of the electrostatic desalting units to assist final emulsion resolution and the removal of brine waters.

This process is critical to the operation of the atmospheric and vacuum distillation processes, since organic salts and water content reduce their distillation efficiency. This process, as in the production facilities plants, results in both a water and oil leg. The oil leg is routed to the distillation processes, while the water leg must be treated for reuse or disposal. The procedures for oil removal from the de-salting wash follow much the same procedures as those employed by the production facilities. The addition of oil in water de-emulsifiers is usually performed on these water legs prior to flocculation and flotation of the suspended oil. Refinery effluent waters are also regulated by various agencies, and because of the volume of discharge produced by these operations, the potential for substantial punitive fines are of major concern.

Emulsion Formation

Crude oil is usually, but not always, associated with water. During the process of its retrieval from the production zone, the produced fluid undergoes a significant amount of agitation. It is this agitation combined with heat, pressure, and chemicals present in the crude that act to produce emulsions. The type of chemicals present in the crude oil are many and varied, and range from pure hydrocarbon (C_nH_{2n+2}) to complex hetero-atomic polycyclics. These also present a range of solubility from water-soluble to oil-soluble, and it is this range of solubilities that is responsible for the formation of emulsions. When a producing well is brought into production, the quantity of water present in the oil is determined by the content of coincident water and oil present in the formation.

Much of the crude oil produced is derived from sandstone formations. These formations consist of combinations of silicon and oxygen that tend to form as partially-charged, anionic (negatively charged) crystallites. These crystallites have a high affinity for water and are often found in close association. This close association is due to the phenomenon of hydrogen bonding, where the partially positive hydrogen of water interacts with the partially negative oxygen of the silicate (Si_nO_{2n}). This interaction and association results in a layer of water surrounding the crystallites, which is termed connate water.

The connate water layer tends to remain closely associated with the silicate surface, and maintains an equilibrium with the free water contained in the crude oil (see Fig. 1–4). Over time this association is established as a static condition, since no external force has acted as an agent to change this preferred state. When the reservoir is tapped, this equilibrium state is disturbed, and the pressure drives the fluid from the pore channels within the sandstone formation. The resulting increase in shearing forces combines with the equilibrium shift of free-water partial pressure in the oil phase, and emulsions begin forming.

Emulsion Formation Criteria

The criteria for the formation of emulsions can be divided into categories:

- Differences in solubility between the continuous phase and the emulsified phase must exist
- Intermediate agents having partial solubility in each of the phases must be present
- Energy sources of the appropriate magnitude to mix the phases must be available

Solubility

The first criterium requires that the phases undergoing emulsification consist of molecules that exhibit wide separations in chemical composition,

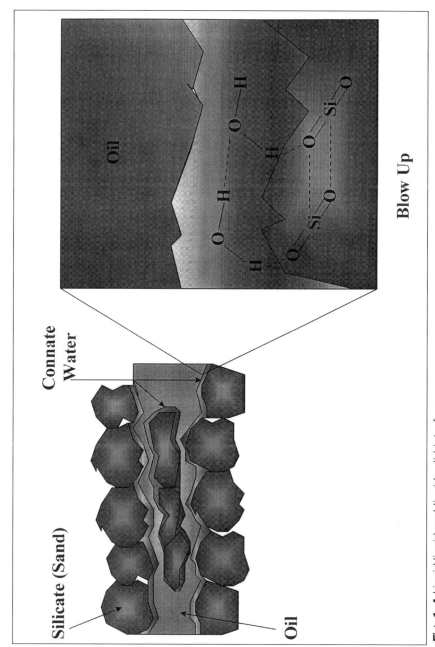

Fig. 1–4 Liquid-liquid and liquid-solid interface

since their compositions determine solubility. A corollary to this requirement is that the physical state of the two phases be the same and that each exist as liquids under prevailing conditions of temperature and pressure. Solubility, as it is commonly thought of, is a term of convenience, since it does not adequately describe the phenomena it represents. Generally, solubility is thought of as a state that exists between a solvent and a solute that exhibits a gross appearance of homogeneity. However, this description is inadequate and represents an observation that is highly subjective.

Solubility should be thought of as the state of division of solute by solvent. This state of division represents a physical condition between the solute and solvent and, as a result of this description, even pure materials can be thought of as solute and solvent. Why make this distinction? The reason is simple: pure systems exist in three physical states—solid, liquid, and gas. These physical states are determined by the prevailing conditions of temperature and pressure, which are manifestations of molecular velocities. The different states, however, are not clearly distinguished by sharp differences of temperature and pressure, but exist in various equilibrium conditions. It is these equilibrium condition values that determine the solubility of a phase within a phase.

The second part of the first criterium for emulsion formation requires that the phases be in the liquid state. This requirement, when applied to pure systems, results in a measure of association between molecules (see Fig. 1–5). Thus, the solid form of the aggregate molecules can be thought of as a physical state that results from the maximum molecular association, the gas as a minimum, and the liquid as an intermediate form of the others. A less obvious but equally important aspect of this description is that molecular aggregate velocities determine the physical states of the system.

Intermediary Agents

The second criteria for the formation of an emulsion requires the presence of an agent possessing partial solubility in both phases. The class of chemicals that represents these agents generally possess functional groups that confer bipolarity to these intermediary molecules. A few examples of bipolar molecules are shown in Figure 1–6.

Although Figure 1–6 shows some good examples of bipolarity, there exists a multitude of molecular variations of this type of bipolarity. However, at this point it is sufficient to point to a class of examples that exhibit this bipolar behavior so the discussion of emulsion criteria can continue.

At equilibrium, the bipolar molecules are aligned as indicated in Figure 1–6 with their nonpolar alkyl (C_nH_{2n+2}) groups extended into the nonpolar oil phase, and their polar groups in the polar phase. This arrangement represents a state of stability that is favored by group interactions and static conditions of temperature and pressure. As these equilibrium arrangements are subjected to shear forces arising from gas or fluid movement they begin forming emulsions. Refer to Appendix A for development of the mixing forces that lead to stable emulsions.

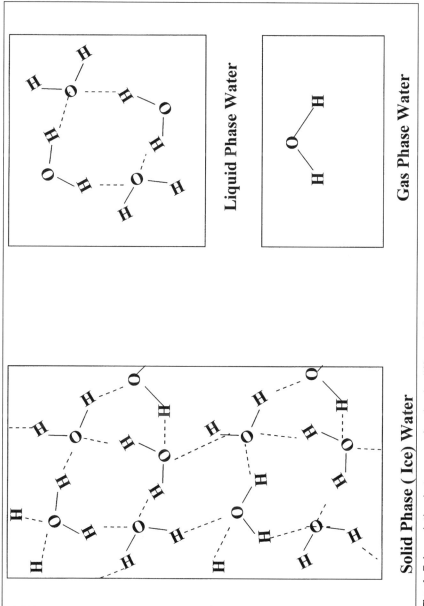

Fig. 1-5 Association between molecules in different phases

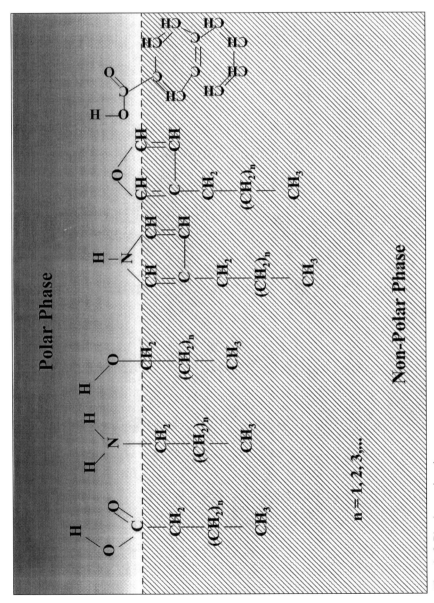

Fig. 1–6 Bipolar molecules

Oil-in-water emulsions exist in various sizes and possess varying degrees of stability. The generally accepted structure of these aggregates is shown in Figure 1–7. In this figure the water/oil interfaces is shown in an expanded view at the left. This expanded view shows the arrangement of the bipolar molecules at the surface between the water and oil. The nonpolar (fatty) tails of the bipolar molecules line up from the oil center with their tails directed into the oil and their polar heads away from the oil center.

The three-layer view shows a second layer of emulsifying molecules that line up polar head to polar head of the first layer. Finally a third layer of bipolar molecules lines up tail to tail with the second layer with their polar heads directed into the polar phase. The polar heads of the third layer are capable of ionic interactions with charged cations (positively charged ion) contained in the polar phase. These cation polar group charge interactions are possible since the polar group can undergo loss of a proton to form a negative anion. Fatty acids ($C_nH_{2n+1}COOH$), fatty alcohols ($C_nH_{2n+1}OH$), and fatty amines ($C_nH_{2n+1}NH_2$) are three such bipolar molecules.

The loss of the hydrogen to the polar phase results in the formation of hydronium ions (H_3O^+), and fatty acid, alcolate, or anionic salts as depicted in Figure 1–8. This then results in the surface presenting an excess negative surface charge (see Fig. 1–9). This excess surface charge is useful, as we shall see later, when devising a chemical treatment to resolve the emulsion.

Water-in-oil emulsions can be viewed as a symmetrically inverted image of the oil-in-water emulsions with layer arrangements as shown in Figure 1–6. This arrangement shows symmetry (see Fig. 1–9) and is consistent with its oil in water analog. The usefulness of this arrangement, in terms of devising treatment chemistries, is less than that of the oil in water because of the inability of the oil phase to conduct charges. The chemicals used to resolve water-in-oil emulsions must either interact by surface penetration or through another mechanism that does not involve electric charge propagation through the nonconductive oil phase.

External and Internal Phase

Since the description of the two emulsion forms (water in oil, and oil in water) represent symmetrically inverted images of the interface, what factors determine the internal and external phase? The strength of the interactions of polar groups of one bipolar group and another, or the polar phase, and interactions of nonpolar groups with one another, or nonpolar phase, helps determine the type of emulsion formed. Thus molecules with strong polar group interaction tend to form water-in-oil emulsions, whereas molecules with weaker polar group interactions tend to form oil-in-water emulsions. Figure 1–10 shows the preferred configuration of sorbitan mono-oleate in a mixture of oil and water. Another factor determining emulsion type is the configuration of the nonpolar group. An example of a bipolar molecule with less hydrogen bonding characteristics than sorbitan mono-oleate is stearic acid. Figure 1–11 shows its preferred arrangement in an oil/water mixture.

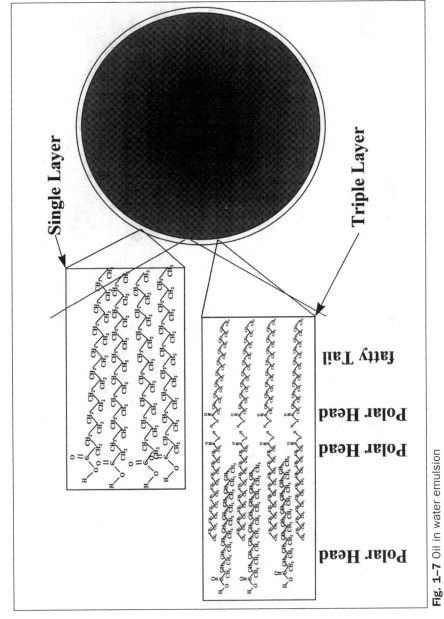

Fig. 1–7 Oil in water emulsion

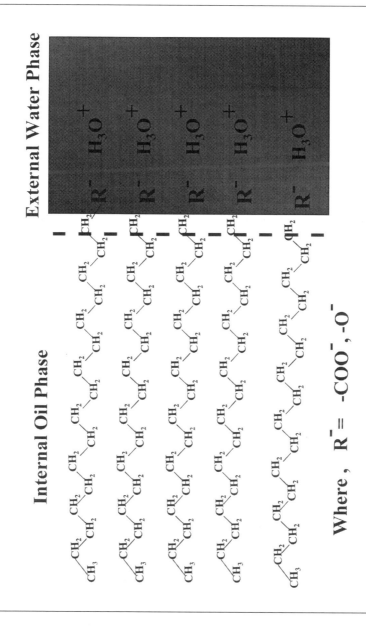

Fig. 1–8 Loss of hydrogen to polar phase

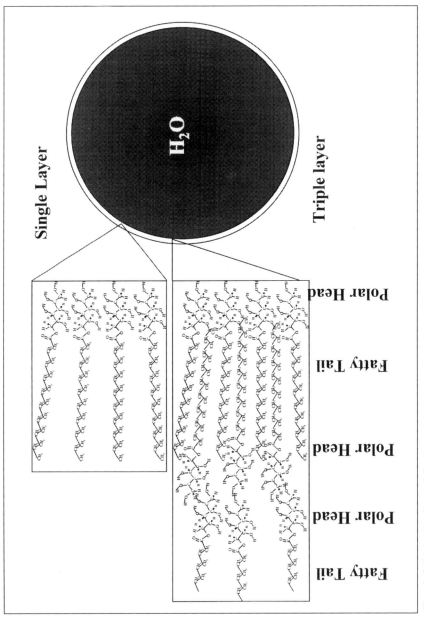

Fig. 1-9 Water-in-oil emulsion

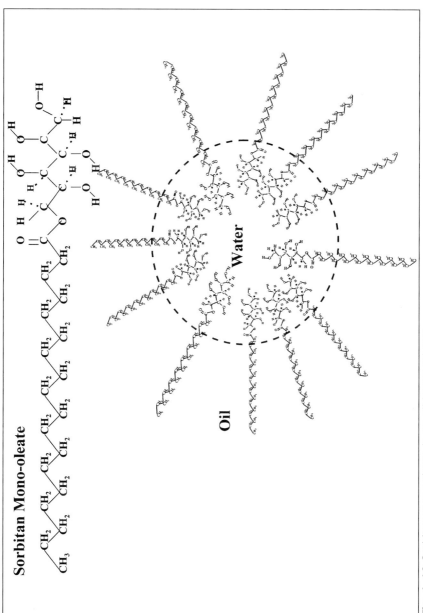

Sorbitan Mono-oleate

Fig. 1–10 Sorbitan mono-oleate in an oil/water mixture

25

If the fatty tail or alkyl group is highly branched, or contains ring structures, the forces of interaction (London dispersion forces) tend to be weaker than those possessed by linear alkyl groups, since orderly arrangement of bulky groups is energetically less favored. Thus, the expected arrangement of naphthoic acid in an oil/water mixture is shown in Figure 1–12. Thus, Figures 1–10, 1–11, and 1–12 show the type of emulsion formed (oil in water, or water in oil) as a result of two main factors: polar interactive forces and molecular structure.

Summary

The preceding chapter discussed the importance of the removal of water from oil, and oil from water to the areas of production, transport, and refining. The key role de-emulsification plays in these areas should have become obvious. Several extremely important points were emphasized in this discussion:

1. Chemical placement and contact time
2. Emulsion settling times
3. Chemical effectiveness
4. Mechanical assistance
5. Environmental conditions
6. Governmental and environmental regulations
7. Chemical effectiveness

Subsequent chapters are aimed at providing an understanding of the chemistry and principles of de-emulsifier technology.

Oil-in-water emulsions (and water-in-oil emulsions) result from the aggregation of several structural subunits. The driving force for the formation of an emulsion is the minimization of surface exposure between two dissimilar phases. Intermediary structures (bipolar molecules) arrange themselves between the two phases, and depending on the strength of the polar interactions and molecular structure, produce water or oil internal structures. The intermediary structures form a three-layer structure that directs their polar heads in for water-internal emulsions and out for oil-internal emulsions.

In equilibrated systems the sizes of emulsions are determined by time, temperature, and emulsion mass. The internal forces of a reservoir are at equilibrium and differential surface structures are optimized prior to being tapped by a well. Once a well is established the dynamics of the reservoir change, and shearing forces result from pressure releases, fluid movement, and temperature changes. This altered dynamic produces emulsions of both types in some cases, and one type in others, depending on the chemistry of the intermediary molecules.

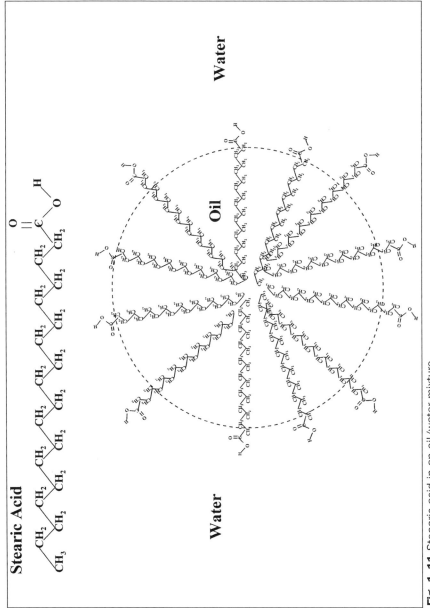

Fig. 1–11 Steaaric acid in an oil/water mixture

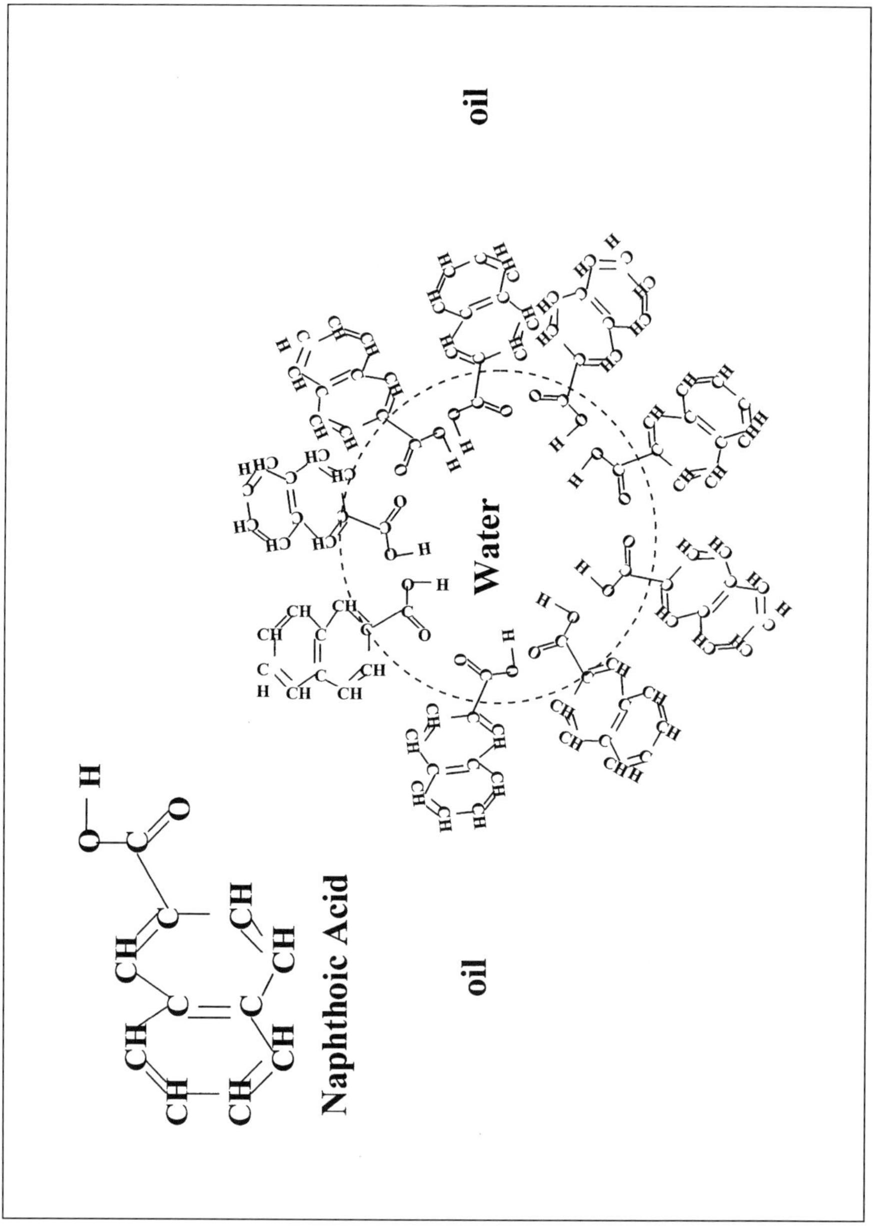

Fig. 1–12
Naphthoic acid in an oil/water mixture

Problems

1–1. What are three criteria for the formation of emulsions? In order of rank, which of these criteria would you expect to be most critical to the stability of the emulsion?

1–2. What characteristics of chemical structures produce bipolarity in a molecule?

1–3. Order the strength of the following polar groups: R-OH, R-NH$_2$, RNHR', COOH.

1–4. What factors influence the size of the emulsion formed?

1–5. What effect does temperature have on the size and stability of an emulsion?

1–6. What type of bonding do alkyl groups of bipolar molecules exhibit?

1–7. What part do gravitational forces play in the stability of emulsions?

References

Mysels, Karol. J. *Introduction to Colloid Chemistry*. 1st ed. New York: Interscience Publishers, Inc., 1959

Handbook of Chemistry and Physics. 56th ed. Cleveland: CRC Press, 1975–1976.

2
Forces Involved in Emulsions

Chemical Complexity of Crude Oil Emulsions

Emulsions are not just pure water in oil, or oil in water, but complex mixtures of a large variety of chemical structures that partition into the phase that provides the most stable environment.

The oil phase is comprised of chemicals that exhibit a wide variety of functionality and structure, as Figure 2–1 shows. Alkanes (C_nH_{2n+2}), mono-alkenes (C_nH_{2n}), aromatic (C_6R_6, where, R= H, alkane, alkene, etc.), hetero-atomic ($C_nR_mR'_{(5-m)}$), where, R_m = O, N, and S, and $R'_{(5-m)}$ = H, alkane, alkene, etc.) are also present in crude oils. This listing of components is by no means exhaustive, but it does illustrate the complex make-up of petroleum crude oil. The situation is not quite as complicated with the polar phase; however, the types of chemicals that favor this environment are also quite numerous. Cations (positively charged ions) and anions (negatively charged ions) comprise the bulk of these chemicals. Acids and bases are included in this category, but for this discussion they will be considered as producing hydronium (H_3O^+) and hydroxyl (OH-) forms from water.

Each of the ions in Table 2–1 are capable of strong ionic interactions with the polar phase of an emulsion system. The ions are hydrated by water, or surrounded by water layers (hydration sheaths). The strength of this hydration sheath is determined by the charge on the ion and its radius. Thus, in Table 2–1 the cations are listed in increasing hydration energy from top to bottom. The anions exhibit the same relationship size to charge and hydra-

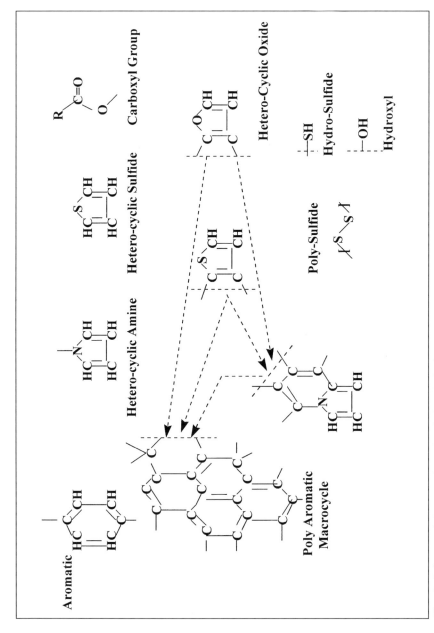

Fig. 2-1 Some of the complex mix of chemicals found in crude oil

tion energy. Therefore given two ions of equal charge but different sizes, the smallest one will exert a stronger force of interaction.

The situation is further complicated by the presence of a few to several other ions in the polar phase. Additionally, the other ions present can be either anionic, cationic, or both. Most of the time, but not always, cations are accompanied by anions. In the case where they are not, the polar phase (water) provides the counter-ion either as a hydronium or a hydroxyl ion. This usually occurs as a result of oxidation of metallic atoms from the surroundings (see Fig. 2–2).

Because of the ease with which water is oxidized, it is not uncommon to find that the anion has been protonated (hydrogen added) to its neutral form (e.g. Cl- to HCl). This gives rise to the hydronium-hydroxyl pair, which now acts to complement the cation and leaves behind the hydroxyl as the new anion. Thus, the polar phase is constantly shifting its equilibrium to maintain its charge stability, and it is this shifting of equilibrium that makes it possible to disturb the balance and resolve the emulsion.

Figure 2–3 shows a representation of the aqueous solvation of iron. The structural groupings of water are represented in two formats: *structure (b)*, where the apex of the dodecahedron has an external hydrogen capable of bonding with a lone pair from an adjacent polyhedron; and *structure (c)*, where the apex presents a lone pair from oxygen that is capable of hydrogen bonding or coordination with a cation. The complexity of the polar phase becomes obvious from these illustrations.

A less obvious feature of the arrangements depicted in Figure 2–3 is that at the periphery of the polar phase, the water will present structure (b) or (c). The presentation of (b) is partially positive, and the presentation of (c) is partially negative. Thus, the interface can exhibit either a positive or a negative charge. Considering this relationship and the structures for water in oil and oil-in-water emulsions depicted in chapter 1, it is possible to understand the reason for a negative charge surface in the oil-in-water emulsion and conversely the positive charge in the water-in-oil emulsion.

Nonpolar Interactions

In the discussion of the polar phase, the forces of attraction are very strong; but when nonpolar interactions are involved, the forces of attraction are considerably less. London forces of induction arise from electron orbital blending, and are caused by the close approach of one atom of a molecule to another atom or molecule. The closeness of this approach is determined by the electronic configuration of the atom or molecule and is restricted to specific radii called the *van der Waals radii*. As like atoms or molecules approach each other, they are repelled by their like charges, but if their velocities and therefore their momentum are sufficient, they can overcome the repulsive force. If they possess just the right momentum they can combine, and the distance between their centers must be a minimum—the van der Waals radius.

Some Common Cations

K^+		
Na^+		
Ba^{++}		
Ca^{++}		
Mn^{++}		
Fe^{++}		
Mg^{++}		
Zn^{++}		
Cu^{++}		
Ni^{++}		
Co^{++}		
Cu^{++}		
Fe^{+++}		
Al^{+++}		

Some Common Anions

Cl^-	OH^-	
I^-	H_2COO^-	
Fl^-	NH_2^-	
NO_3^-	HS^-	
$H_2PO_4^-$	HSO_4^-	
	CN^-	
HPO_4^{--}	$HCOO^{--}$	
BO_4^{--}	SO_4^{--}	
PO_4^{---}		

Table 2–1 Common cations and anions

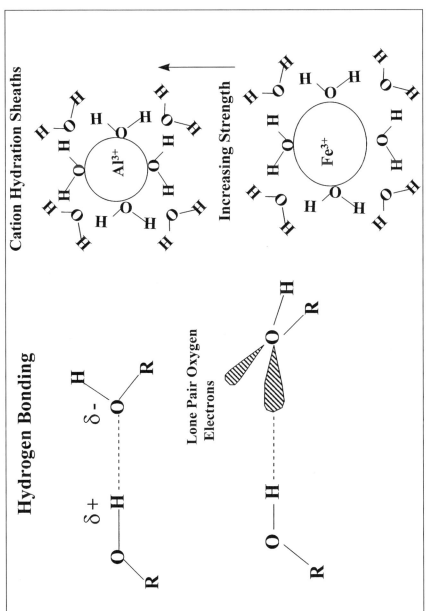

Fig. 2-2 Polar phase hydronium/hydroxyl equilibrium

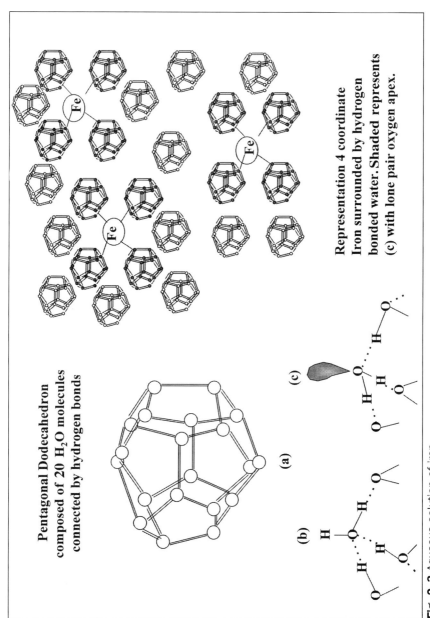

Pentagonal Dodecahedron composed of 20 H₂O molecules connected by hydrogen bonds

Representation 4 coordinate Iron surrounded by hydrogen bonded water. Shaded represents (c) with lone pair oxygen apex.

Fig. 2–3 Aqueous solution of iron

At the van der Waals radius, the electron clouds interchange electrons, and one atom or molecule takes on a partial negative or positive charge while its conjugate atom or molecule takes on its opposite charge. Figure 2–4 shows a schematic representation of the London forces, and particular note should be taken of the fact that these forces occur in molecules possessing no permanent dipole. These forces are weak, as mentioned above, and in order to contribute significantly to the strength of a macro structural arrangement, they must be numerous. Therefore, the structure of the molecules involved in these inductive interactions must afford each other sufficient room to approach each other to a distance equal to the van der Waals radius.

When molecules are highly branched, or contain bulky substituents, the effective inductive interactions are reduced. This reduction in effective inductive interactions results in a less stable aggregate structure. Thus, it is clear why the structure of naphthoic acid emulsion is less stable than that of sorbitan mono-oleate as shown in Figure 2–5.

Neither structure depicted in Figure 2–5 shows the full complement of second- or third-layer bipolar emulsifiers. However, it should be seen that sorbitan mono-oleate can accommodate more alkyl units in each successive layer than naphthoic groups can accept additional naphthoic groups. Figure 2–6 indicates the inductive charge arrangements between two nearby alkyl groups that appear in Figure 2–5 in the sorbitan mono-oleate emulsion.

Combined Ionic and Inductive Forces

The combination of ionic and inductive forces produces emulsions of varying stability, and from the argument presented in Appendix A we can see that the stability increases in inverse ratio to the size of the emulsion. Upon examination of the sorbitan mono-oleate emulsion versus the naphthoic emulsion, it is the bipolar emulsifier that has the least hindrance to inductive interaction that will also form the smallest emulsion. Additionally, the degree to which hydrogen bonding can occur between the emulsifier and the components of the polar phase is a function of the size of the nonpolar portion of the emulsifier. The complex ionic nature of the polar phase of an emulsion is evident when an examination of the possible combinations of cations, anions, and hydration sheathes are contemplated. The number of permutations of the cations and anions that can be present in the polar phase is extremely large. Likewise, the possible organic structures that may be present in the nonpolar phase are immense when alkyl substituents are added to the structures shown in Figure 2–1.

Aggregate Interactions

The preceding discussions have developed a picture of emulsions that conveys the extremely complex nature of these aggregate structures. This complexity arises from several contributing factors, among which are the immense variety of molecular structure, ionic and inductive interactions, sin-

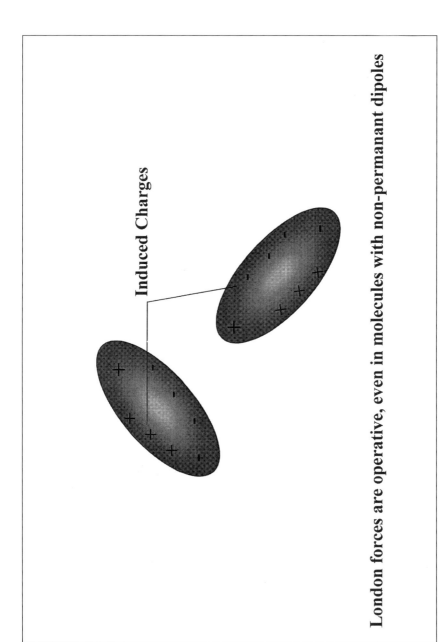

Fig. 2–4 London forces

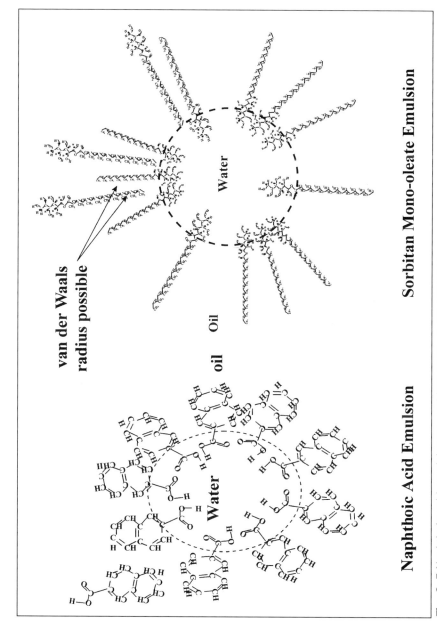

Fig. 2–5 Naphthoic acid emulsion vs. sorbitan mono-oleate emulsion

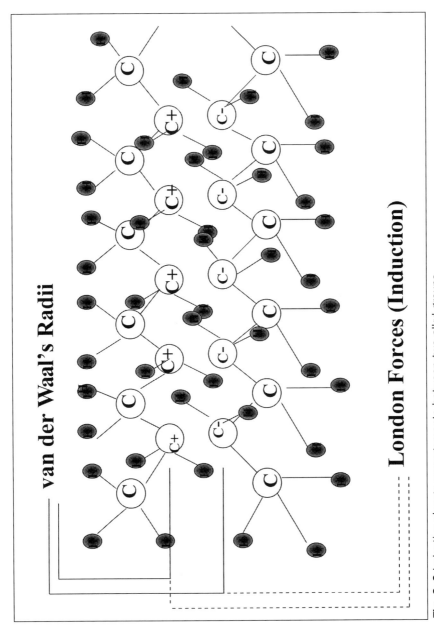

Fig. 2–6 Inductive charge arrangements between two alkyl groups

gle and multiple bipolar layering, hydrogen bonding, solvent sheathing, metal coordinate complexes, and charge sign at the interface. Additional complications arise when the physical states and interactions of the macro-aggregates are considered. Some of these interactions include partitioning of the bipolar phases, diffusion of polar phase between aggregates, aggregate number, aggregate collision frequency, collision energy, gravitational settling, and surface tension.

Bipolar Partitioning

The bipolar emulsifiers present in a biphased system will partition into collections of like species or molecular structure. This occurs because various molecules exhibit different behaviors under different conditions of temperature and pressure. Two of the bipolar molecules mentioned earlier (naphthoic and stearic acids) provide good examples of these different behaviors. The melting points of stearic and naphthoic acids are 71.5° C and 185.5° C, respectively. Both of these acids are found in crude oil, and therefore represent good candidates for discussion. Although the hydrogen bonding capabilities of naphthoic acid are limited, aromatic ring interactions of the unpaired electrons plus the carboxyl group interactions combine to produce its high boiling point. Solvation by the nonpolar phase is therefore less successful than it is in the case of stearic acid.

Additionally, the interactions of the carboxyl and aromatic substituents provide a much more stable aggregate than the stearic acid's inductive alkyl and carboxyl interactions. Thus, the naphthoic and stearic acids will tend to aggregate in groups of like molecules. These aggregate groupings will collect at the interface between the nonpolar and polar phases and remain grouped at the interface (see Fig. 2–7). This explains why the emulsions formed in a mixed system tend to exhibit a partitioning of bipolar emulsifying agents.

These partitioned groupings, however, do not necessarily produce smaller, stronger emulsions simply because of their intermolecular attraction forces. The strength of the intermolecular attractions must be overcome, to some degree, when the ordered emulsifier layer is formed. Thus, the geometries of the groupings, or molecular positions, are altered in going from one orientation to another, and mixed-phase emulsifier systems tend to produce emulsions with sizes that reflect the various bipolar phases present in the system.

Internal Phase Diffusion

Emulsions can be thought of as containers for chemically dissimilar materials occupying space inside a continuous phase of opposite polarity. These containers are semipermeable and allow interchanges of similar and appropriately sized fractions. In this way a dynamic equilibrium is set up between containers (emulsion aggregates) that maintains a balanced concentration of internal phase solutions within similarly composed aggregates. Thus, an emulsion formed from a highly concentrated ionic water

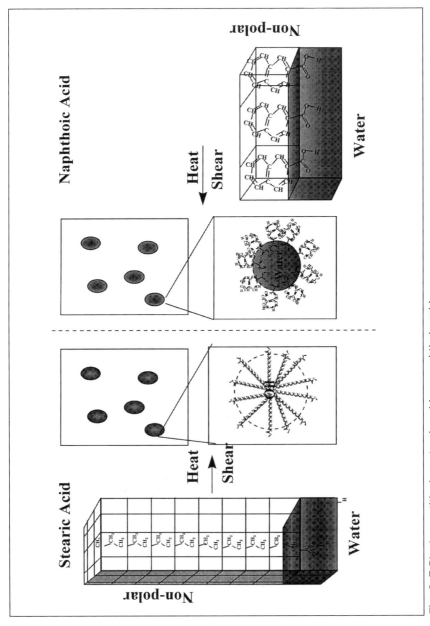

Fig. 2–7 Bipolar partitioning: stearic acid vs. naphthoic acid

solution, oil, and naphthoic acid will equilibrate concentration imbalances resulting from its formation by the diffusion of water to or from the emulsion aggregate.

The same is true for the nonpolar phase and an equilibrium concentration of solutes within the internal nonpolar phase of an emulsion is also maintained by diffusion. However crude oils seldom consist of a single bipolar emulsifier; consequently, the internal phase solute concentrations of different emulsion forms also differ. Since these concentrations differ, the size of the emulsion aggregates contained in the oil or water comprise a range of sizes and therefore a range of differing stability. Although it is often true that a system will exhibit a predominate emulsion form, for example water in oil or oil in water, the existence of both forms in the same system is not precluded. In fact it is a safe bet that in crude oil systems the existence of both forms is probably the rule rather than the exception, considering the variety of bipolar species present.

Aggregate Number

The number of emulsion aggregates present in a biphased system is a function of

- their size
- the amount of polar and nonpolar phase
- the number of bipolar emulsifier molecules present
- the amount of shearing energy applied to the system
- the viscosity of each phase
- the temperature and pressure of the system

Each emulsion aggregate formed by a specific emulsifier has a critical number of bipolar molecules required for the formation of a minimized surface. This number is then a function of the geometric surface produced, and the geometric arrangement of the emulsifier required for polar and nonpolar forces of attraction to become effective. Thus, a rough number can be calculated that takes into account the spacial displacement of the bipolar molecule at the surface.

An example of such a calculation would be as follows: The calculated cross sectional displacement of a stearic acid molecule is near 20×10^{-16} cm^2. If an emulsion has a radius of 1 micron or 10^{-4} cm and the emulsion is spherical, then the emulsion surface area is $2\pi r^2 \approx 6.28 \times 10^{-8}$ cm^2. The emulsion surface area can then accommodate $\approx (6.28 \times 10^{-8} / 2 \times 10^{-17}) \approx 3.14 \times 10^9$ molecules of stearic acid. All the emulsions are of the same size, and there is 10 cm^3 water and 400 cm^3 of oil involved in emulsion. The total number of identical 1 micron (m) radius spheres of water would be $\approx \{10 \text{ cm}^3/(4\text{p}r^3/3)\} \approx 2.38 \times 10^{13}$ or $(2.38 \times 10^{13})/(4 \times 10^2) \approx 5.6 \times 10^{11}$ emulsion aggregates per cubic centimeter of the system. These calculations yield approximately 7.9×10^{11} grams (g) of water and 1.5×10^{18} g stearic acid per emulsion aggregate.

Emulsion Collision Frequency

It is possible to calculate a rough estimate of the number of collisions an emulsion system undergoes per second per cubic centimeter by employing the collision theory for gases in combination with Boltzmanns' distribution. The beginnings of this approach involve the determination of the mean free path, or distance traveled between collisions, and is illustrated in Figure 2–8. This figure shows the path swept out by sphere A in one second with average velocity v, and a diameter of s. The first thing to do is to investigate the number of collisions per second that a sphere makes. This is called the collision number and is denoted by the following:

$$Z_1 = \sqrt{2\,\pi\sigma^2 v N^*}$$

The next thing to do is determine the number of collisions that will occur per unit volume per unit of time. According to Gordon Barrow, *Physical Chemistry*, "the collision number is closely related to Z_1 since there are N^* molecules per cubic centimeter and each of these molecules has Z_1 collisions per second, the total number of collisions per second per cubic centimeter will be $1/2\, N^*\, Z_1$." This leads to the following result for the collision frequency:

$$Z_{11} = 1/\sqrt{2\,\pi o^2 v\, (N^*)^2}$$

In the case of the stearic acid emulsion described in the section on aggregate number, the radius of the emulsion was given as 1 micron (μ), thus the diameter σ for the emulsion is 2μ. This leaves the average velocity v as the unknown, but we know the number of emulsion aggregates is 5.6×10^{11} aggregates per cubic centimeter of system. Thus using Boltzmann's distribution, an average velocity can be determined ($T = 298°$ K) the following gives a value for the average velocity of the emulsion aggregates.

$$v = (3\ nkT/m)^{1/2}$$

where

k is 1.38×10^{-23} joules/ (molec. ° K)
n is equal to $N^* = 5.6 \times 10^{11}$
m is 7.9×10^{-11} grams

Going through the calculations, an average aggregate velocity v is calculated to be ≈ 9.4 meters/second (m/sec). Substituting this average velocity into the calculation for Z_{11}, the total value of collisions per cubic centimeter per second is found to be 2.6×10^{27}. This value is an order of magnitude less than that calculated for nitrogen gas at the same temperature and pressure ($298°$ K and 1 atmosphere), where Z_{11} for nitrogen is 8.99×10^{28}. Thus, the collision frequency of an emulsion is quite low even

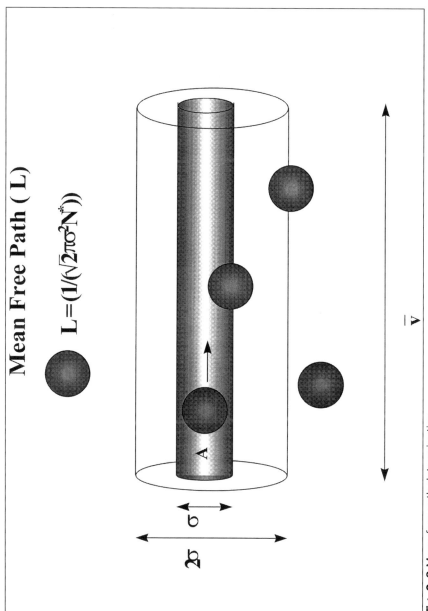

Fig. 2-8 Mean free-path determination

compared to the collision frequency of a gas under standard conditions of temperature and pressure.

Emulsion Collision Energy

The collision frequency between emulsions has been determined to be quite low at room temperature and 1 atmosphere (atm) of pressure. Therefore it would be expected that the frequency of emulsion entities combining to form larger emulsions would also be relatively infrequent. However, in order to determine the frequency of combination we must consider the energy required for the combination to occur. The average velocity of the stearic acid emulsion was calculated to be 9.35 m/sec, and in order to calculate an energy we must use the following equation:

$$K.E. = (mv^2/2)$$

where

m is the grams/ emulsion aggregate = 7.9×10^{-11} g
v is 935 cm/second.

Calculating through, K.E. = 6.9×10^5 ergs (gm cm^2/sec^2) per aggregate emulsion or 6.9×10^5 dynes cm^2. From surface tension measurements of a monomolecular layer of stearic acid adsorbed on water in air, a value of 43 dynes/cm has been determined. If this value is multiplied by the emulsion volume, then an energy value for effective combination should result. The result of this calculation turns out to be $\approx 1.8 \times 10^{-12}$ dynes cm^2.

Thus, the energy necessary for combination is exceeded by a factor of nearly 10^8 for single layer of stearic acid surrounding an emulsified sphere of water with a 1 μ radius. This indicates that on average each collision between emulsion aggregates will have more than sufficient energy to combine. Therefore it is the relatively infrequent number of collisions, and the number of lower velocity aggregates of the Gaussian velocity distribution, that prevent the spontaneous resolution of the emulsion.

Gravitational Settling Forces

Rather than going through any more calculations, it may be instructive to consider the value for the average velocity of the aggregate emulsion obtained above in relation to the acceleration due to gravity (g). Integrating g with respect to time (using 1 sec), a velocity vector pointing down is obtained with a magnitude of 980 cm/sec. Considering the average velocity of the aggregates is 935 cm/sec, any vector product of these two velocities will result in a downward direction. Thus, the system of aggregate emulsion would be expected to settle in the earth's gravitational field. It is the relatively infrequent number of collisions, and the number of higher velocity aggregates of the Gaussian velocity distribution, that prevent the rapid settling of the emulsion.

Summary

The preceding chapter shows the complex ionic and inductive forces of interaction that are present in emulsions. The ordered arrangement of each component including the ions, water, and emulsifier is determined by the interactive forces between the components. And the strength of these interactions is determined by the structure and number of subunits involved. Although no specific reference to the effects of velocity were made, the introduction of London inductive forces and the van der Waals radii were included to show its implicit effects. These implicit effects center around the need for a minimum approach distance, achievable only by appropriate momenta that result from molecular velocity.

This chapter also discussed the characteristics of a stearic acid, water, and nonpolar external phase emulsion. A hypothetical uniform-sized (1 micron) emulsion was developed consisting of a single layer of bipolar emulsifier ($\approx 8.4 \times 10^7$ g of stearic acid was required), 10 cm^3 of water, and 400 cm^3 of nonpolar phase. Calculations were performed to determine aggregate number, mean free path, collision frequency, required coalescence (combinational) energies, and settling tendency. As a result of these calculations, it was found that although a single layer bipolar emulsifier could be constructed from the quantities supplied, that this emulsion would not be stable.

What the calculations did not determine was the ability of the emulsifier to form larger aggregate emulsions, which would exclude some of the water phase. Neither did they show the effects of multilayering of the emulsifier and similar exclusion of some of the water phase. Perhaps the most important aspect is the insight into the semiquantitative characteristics of emulsion systems. The relatively infrequent collisions of emulsion aggregates, their low average velocities, their surface tension, and their energies tend to come into focus when exercises like the one above are performed.

Problems

2–1. What factors determine the tendency of like molecules to partition into groups when several bipolar molecules are present?

2–2. Would you expect several different emulsion sizes in a given system? If so, why? If not, why?

2–3. Given s = 10^{-10} meters, and $N^* = 6.023 \times 10^{23}$, calculate the mean free path L.

2–4. If the velocity of the aggregate is 420 m/sec, calculate the collision frequency Z_{11}.

2–5. What characteristic of emulsion aggregates can be seen in gas phase collision phenomena?

References

Barrow, Gordon M. *Physical Chemistry*. 2nd ed. New York: McGraw-Hill Book Co., 1966.

Handbook of Chemistry and Physics. 56th ed. Cleveland: CRC Press, 1975–1976.

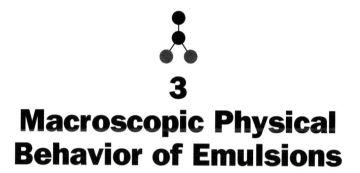

3
Macroscopic Physical Behavior of Emulsions

Viscosity

Viscosity effects exhibited by emulsions tend to be complicated due to their variable composition, average emulsion size, and the individual viscosities of their phases. Thus, measurements of viscosity over widely varying conditions of temperature and shear stress are necessary to elucidate the possible interactions responsible. Consequently, dynamic instruments like the one depicted in Figure 3–1 are most valuable.

The dynamic viscometer uses a rotary spindle of known surface area that is attached to a strain gauge of some construction (e.g., spring, potentiometer, or piezoelectric) that measures the resistance to a known torque force. Viscosity can be described as a measure of that energy dissipated by a fluid in motion as it resists an applied shearing force. Because the resistance is a function of the difference in velocities of infinitesimally small layers (lamella) of unit area, then the summation of the lamella interactive forces is termed viscosity and given the units of poise (gram/cm-second).

Since the interactive forces involved in emulsions are quite complex, it is appropriate to describe the gross behavior of these systems first, and, if possible, later try to account for the behavior. The viscosity profile of a single-phase system (e.g., glycerin) can be obtained under varying conditions of applied shear, changing temperature, changing pressure, or combinations thereof. The general behavior of the glycerin under a constantly increasing

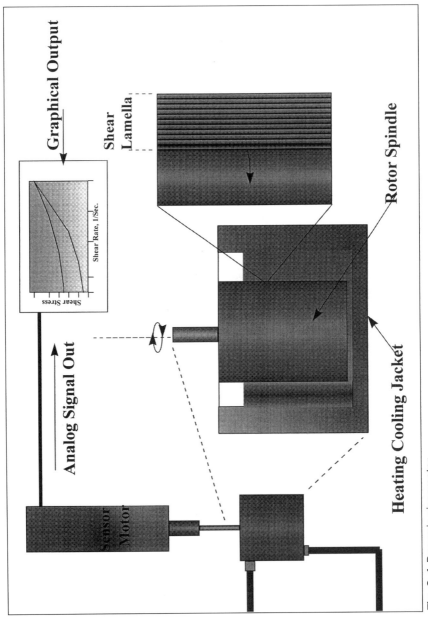

Fig. 3-1 Dynamic viscometer

applied shear stress is said to be Newtonian. Newtonian viscosity is best described graphically, but it means that glycerin's resistance to a constantly changing applied shear stress (torque for spindle) increases as the shear stress increases and retraces the same curve when the shear stress rate change is reversed.

Another viscosity profile behavior is classified as non-Newtonian, but this category comprises two major sub-categories, dilatant and thixotropic. Dilatant materials are not all that common, but a good example of this behavior is exhibited by poly-methyl-methacrylate (motor oil additive). The poly-methyl-methacrylate polymer undergoes shear thickening when a shearing force is applied, thus giving the motor oil greater lubrication properties at higher shear (often sold as SAE 10W-30). Thixotropy is a more commonly exhibited behavior, and represents the thinning of a fluid as it undergoes a shear stress. Examples of thixotropy are the thinning of catsup when shaken, latex paint phase shearing, and shear thinning of wax solutions. Figure 3–2 shows examples of each viscosity profile.

Figure 3–2 lists the hydrogen bonding forces of glycerin as exhibiting a Newtonian viscosity profile because under shear stress these forces resist elastically (hydrogen bonds are not destroyed, only distorted). The wax solution is listed as thixotropic because the London forces of induction between wax molecules are broken by the shear stress. Finally, the poly-methyl-methacrylate is listed as dilatant because the high molecular weight polymer is tangled between shearing planes, and no bonds are broken. The preceding discussion was intended to help develop an intuitive understanding of the types of viscosity profiles seen in emulsion systems rather than the rigorous mathematical development of the subject. Thus, viscosity profiles are measurements of the summation of the individual interactive forces between molecules, as well as their composite macroscopic aggregations.

Emulsion Behavior under Shear Stress

Emulsions are discretely segregated fractions of either polar or nonpolar phases that are dispersed throughout a system of opposite polarity. These discrete entities contain fluids that exhibit fluid characteristics of density, viscosity, melting point, boiling point, vapor pressure, and surface tension that can be similar to or vastly different than the surrounding phase. Each emulsion aggregate, therefore, presents itself as a discontinuity within a system. The extent to which the internal phase fluid characteristics differ from the external phase and the number of the discontinuities determine the resistance to shearing forces the system will exhibit.

It is fortunate that in crude oil emulsion systems the polar phase is usually water containing various concentrations of ions, and it exhibits fluid characteristics closely related to those of water. The same is not true of the nonpolar phase or oil-phase fractions. These fractions generally do exhibit widely varying fluid properties. Sometimes a light crude oil with a viscosity less than water contains a water emulsion, and the emulsion aggregates greatly

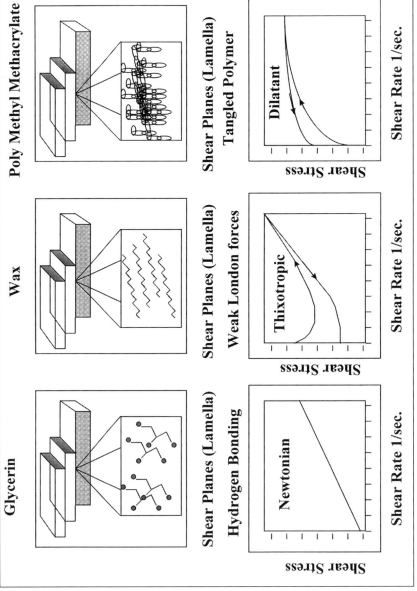

Fig. 3–2 Viscosity profiles

increase the shear stress profile of the system. At another time a heavy crude oil with a viscosity that is much greater than water contains a water emulsion, and the effect of the emulsion aggregates is the reduction of the system shear stress profile. In both cases, the shear stress profile, when plotted against shear rate, produces curves that deviate from smoothness, and the magnitude of these deviations is a measure of the effects of the emulsion aggregates.

Two factors determining whether or not these systems exhibit dilatant or thixotropic behavior include the concentrations of bipolar emulsifiers present at the interface and the equilibrium vapor pressure of the internal phase. In the case of bipolar emulsifier concentration, a possibility exists for a multilayer emulsion to redistribute the bipolar emulsifier when the system is subjected to shearing forces. This redistribution could result in a larger average emulsion size, and therefore a higher system viscosity. In the case of equilibrium vapor pressure, the shearing forces that produce larger average emulsion size may temporarily shift the internal phase component concentration and produce a diffusion gradient between emulsion aggregates.

Temperature Effects on Emulsions

Temperature effects on emulsion aggregates are quite dramatic since the partial pressures of the internal phase solvents increase or decrease in response to temperature shifts up or down respectively. It is important to remember that the emulsion aggregates comprise separate vesicles (containers) for solute solvent systems that exhibit polarity differences from the external phase. The stability of these emulsions is dependent upon an optimized solvent solute concentration, and the vapor pressure is key to the maintenance of that stability.

Under conditions of constant temperature and pressure, emulsion systems maintain a continuous equilibrium transfer of internal phase solvent between emulsion aggregates. When the temperature is increased, this equilibrium balance is shifted, and a more rapid interchange occurs. If the temperature becomes great enough, the solute molecules being transferred will obtain sufficient energy to escape the system, and the internal phase will be depleted of solvent. Meanwhile the external phase is also being depleted of solvent, and the differential rate of solvent loss between the two phases determines the overall system stability.

The dynamics of this temperature effect on emulsion aggregates are depicted in Figure 3–3. Note that there exists a potential nonemulsified reservoir for the effective combination of solvent molecules at the interface between the emulsion aggregates and at the interface between the external phase and air. These reservoirs are extremely unstable since they are not surrounded by bipolar emulsifiers, and they can act as intermediary solvent sinks or as a coalescing locus. The larger these reservoirs become, the less stable they become. Therefore, when the temperature is increased, the solvent collision frequencies and energies increase, and the size of the reservoirs increases. Meanwhile, the external phase vapor pressure (molecular velocities)

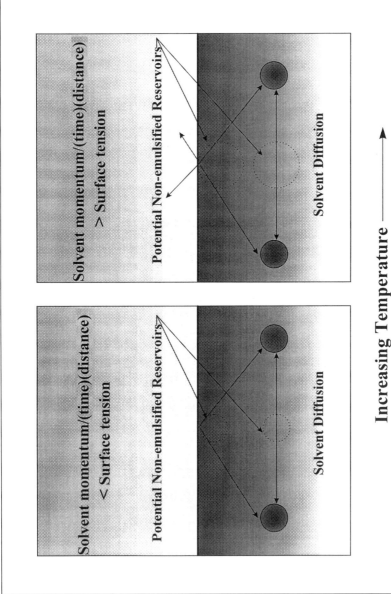

Fig. 3–3 Temperature effects on emulsion dynamics

increases until the molecular momentum expressed over a distance per given time period is sufficient to overcome the surface tension of the system.

The combined result is an increase in the relative size of the non-emulsified reservoirs, a decrease in the size of the emulsion aggregates, and reduction in the volume of the external phase. This process then seems to suggest that the effects of temperature increases result in a tighter emulsion (smaller) and higher exchange rates of solvent. This development might be true if it were not for the action of other external force agents, such as gravity.

Gravitational Effects on Emulsions

The temperature effects discussed above seem to lead to either solvent-depleted systems or tighter (smaller) emulsions, when considered alone, but gravitational factors are overlooked in this discussion. An important derivation leads to an expression for the effects of gravity on emulsion aggregate systems. This derivation is known as *Stokes' Law*, and it so vital to the understanding of emulsion behavior. According to the *Handbook of Chemistry and Physics*:

> Stokes's law.-1. Gives the rate of fall of a small sphere in a viscous fluid. When a small sphere falls under the action of gravity through a viscous medium it ultimately acquires a constant velocity,

$$V = 2ga^2(d_1 - d_2)/ 9\eta$$

> where

> a is the radius of the sphere
> d_1 and d_2 are the densities of the sphere and the medium, respectively
> η is the coefficient of viscosity

> V will be in cm/sec if g is in cm/sec^2, a in cm, d_1 and d_2 in g/cm^3, and η in dyne-sec/cm^2 or poises.

The combination of this law and the discussions of the effects of temperature clarify the behavior of an aggregate emulsion system. When the temperature is raised the sizes of the solvent reservoirs increase, and the emulsion sizes decrease. The reservoirs are pure solvent, while the emulsions are combinations of solute and solvent. The reservoir's density is therefore less than that of the emulsion, and consequently the ratio of phase density (polar to nonpolar phase density) is also less.

Thus, in accordance with Stokes' law, it is expected that the differential densities ($d_1 - d_2$) will oppose a settling velocity (V) increase. However, since the reservoir's radius is increased, the a^2 term in the equation will have a more dramatic effect on the settling velocity V than the differential densities. Thus,

it can be seen that the dynamics of diffusion and the effects of gravity and temperature combine to rearrange phases.

Finally, the viscosity term in Stokes' equation can be examined in terms of the effect of temperature. A molecular kinetic theory expression for the behavior of viscosity with temperature is given by the following expression:

$$\eta = Ae^{\Delta Evisc/RT}$$

where

A is some undetermined constant

$\Delta Evisc$ is the change in viscosity activation energy

Thus it can be seen that as the temperature increases, the viscosity is exponentially lowered, which leads to an increase in the settling velocity V in accordance with Stokes' Law (see Fig. 3–4).

Electromagnetic Field Effects on Emulsions

Emulsions are susceptible to forces exerted by electric fields, but the strength of the electric field effects are dependent upon the external phase's ability to conduct electricity. Thus, a polar external phase emulsion is more susceptible to an externally applied current than an externally nonpolar emulsion. From the description of oil/internal water/external emulsions like the stearic acid system, an overall negative emulsion surface is expected.

When an electric current is supplied to a stearic acid emulsion of oil in water, a migration of the counterbalancing cation to the anode (negative pole) and the anionic emulsion aggregate to the cathode (positive pole) is expected. This migration is possible because the water phase is conductive to electricity. However, in the case of the water-in-oil emulsion the nonpolar oil external phase is nearly an insulator since its resistivity is extremely high. Therefore, the migration of ions as a result of current flowing through the external phase is minimized.

Although current is not conducted to any significant degree through an oil external emulsion, the electric field is still present between the capacitor plates depicted in Figure 3–5. When a changing field force vector is imposed across the capacitor plates as it is with an alternating current source, the relatively small cations in the water-in-oil emulsion are caused to vibrate more rapidly than the larger anions. This displacement from an equilibrium position causes a deformation of the emulsion, and a destabilization of the interface surface.

Figure 3–6 shows the effects of a high alternating current on the shape of an oil external emulsion. Note that the distortion caused by a frequently changing electric vector acts to distort the spherical geometry (minimized emulsion surface), and by doing so destabilizes the emulsion surface. Refer

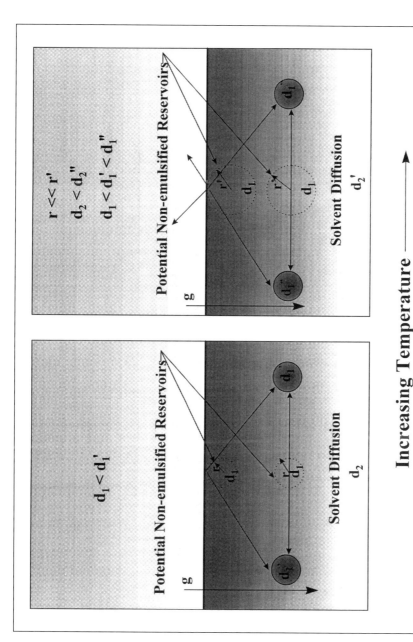

Fig. 3–4 Stokes' law and temperature effects

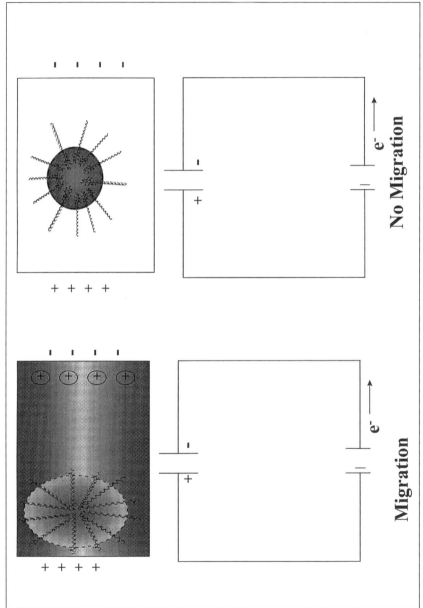

Fig. 3–5 Migration of ions in an electric field

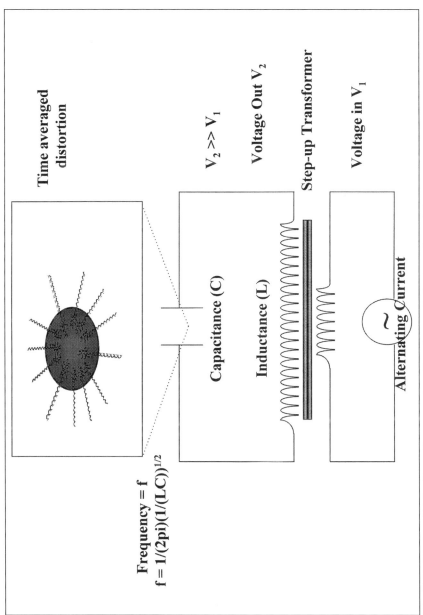

Fig. 3–6 Effects of high alternating current on the shape of oil-external emulsion

to appendix B for a development of the effects of alternating current on water-in-oil emulsions and the relationship of these effects to the natural charge surface.

Determination of Emulsion Type

This section will be concerned with some of the methods used to test emulsions in crude oil systems. Sometimes the most reliable methods are also the most simple, and fortunately this is true in the case of determination of the type of emulsions present in crude oils. Knowing the character of an emulsion is the key to knowing what methods to use in their resolution; therefore, the following test is employed to make this determination.

A simple test for determining whether an emulsion is oil or water external consists of placing a small portion of the sample into two beakers, one containing water, and the other containing xylene, and stirring the contents. If the emulsion is oil external, it will disperse readily in the xylene; if the emulsion is water external, it will disperse readily in the water.

Quantifying Phases

Once the type emulsion is determined, it is necessary to make a determination of the amount of internal phase present in the emulsion. Here again the most simple and direct approach is usually the best approach. The general method used is to dilute the emulsion in a solvent that has been found to disperse it as in the type determination used above. Usually the emulsion is dispersed in a 2:1 ratio of solvent to emulsion, or higher ratio if necessary. Once the emulsion has been diluted, it is subjected to heating and centrifuged to separate the phases. This process usually employs graduated centrifuge tubes, and readings are taken directly of the phase separation (see Fig. 3–7).

Often the emulsion is particularly stable and heated centrifugation is insufficient to resolve the phases, so a de-emulsifier is added to the mixture. This de-emulsifier usually consists of a mixed combination of alkyl-substituted phenolic resin oxyalkylated with ethylene oxide, and an amine sulfonate (this chemistry will be discussed later). This analysis is typically called a *grind out* and sets the bench mark criteria for chemical treatment effectiveness.

Quantifying Phases: Oil-in-Water Emulsions

Although the procedure for determining the quantity of water-in-oil emulsions could be applied to the quantity of oil in water, it is not found to be an effective means of measurement. Thus other methods are employed for this determination. When time allows, a Soxlet extraction procedure provides an effective method for this determination. This method involves the placement of a sample of the oil-in-water emulsion in a specially designed

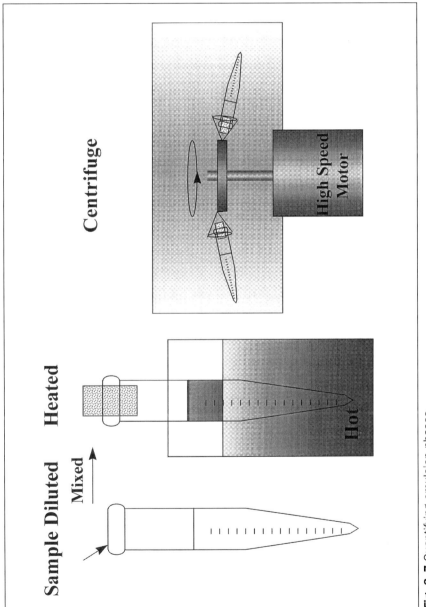

Fig. 3–7 Quantifying emulsion phases

glass adapter that is placed between a reflux flask and a water cooled condenser (see Figure 3–8).

The sample is placed in a compressed cellulose thimble (filter), and the hexane is heated to reflux and cycled through the sample. Once the extraction has cycled several times, the hexane solvent is boiled off, and the residual oils are quantified by weight. This test is very accurate, but it is time-consuming.

Additional test procedure include freon cold wash extractions, where the sample is placed in a separatory funnel containing freon and agitated vigorously to extract the oil. Where possible (available clean, dry oil), a standard curve of light absorption versus concentration is plotted as a standard. Samples of the emulsion are then washed in the freon, and their adsorption spectra compared to the standard curve. This test procedure has likely been discontinued because of environmental concerns, or the freon has been replaced with a less damaging (to the atmospheric ozone layer) solvent.

Product Screening Methods

Testing procedures for determining the effectiveness of water in oil de-emulsifiers have not varied much over the years; they remain very nearly the same as testing procedures conducted 30 to 40 years ago. This procedure is to acquire a sample of oil from the field, agitate it vigorously for about 30 minutes, and pour samples into several graduated pyrex glass tubes (about 100 cm^3). Using microliter dispensers, a concentration range of de-emulsifier chemicals are added, and the tubes are then placed in a heating bath and heated to the temperature to which the oil is exposed in the field.

After heating, the samples are placed on a shaking device, which gently agitates the contents for a standard period of time (about 5 minutes). The samples are again placed in the heating bath, and periodically observed for water drop. As mentioned above under quantifying procedures for water-in-oil emulsions, the benchmark has been set for the treatment effectiveness. Thus, the heating is usually continued until a good approximation of this bench mark value is reached by at least one of the chemicals being screened. Once this approximate bench mark target has been reached, syringed samples are retrieved at various levels of the samples (usually, the top 1/3, middle, and bottom 1/3) and subjected to the quantifying procedure described above.

Oil-in-Water Emulsion Screening

The screening procedures for oil-in-water emulsions are usually conducted in clear glass sample bottles. A 100 cm^3 sample of the emulsion is injected with microliter quantities of oil in water de-emulsifier chemicals, and the samples are gently hand-shaken for 1–3 minutes. The samples are then visually compared using either an artificial backlight or sunlight. The samples that give the clearest transmission of light are judged the best. In the case where chemical treatments yield close comparisons, the freon extraction

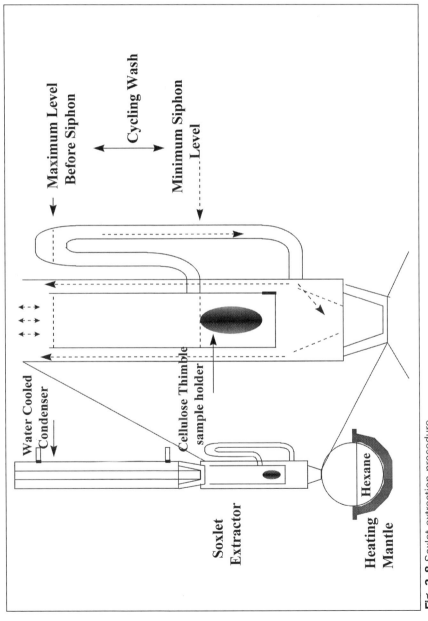

Fig. 3-8 Soxlet extraction procedure

procedure is conducted, and oil concentration is compared to standard adsorption curves.

Although the testing procedures mentioned previously for both oil in water and water-in-oil emulsions appear to be very simple, they are surprisingly accurate and reproducible. This is why uniformity of test methodologies appears to be universal from one specialty chemical company to another. This is not to say that the research being conducted is uniform between these companies. Much research effort has been expended in the elucidation of the mechanisms involved in emulsion resolution. This research requires that a more thorough test scenario be utilized than those listed above; however, these methods should also be included in the testing.

Special Test Procedures

Due to the use of desalting units in refineries, special test equipment and testing procedures have been devised for screening chemical de-emulsifiers in desalting applications. One such device is a device called EDA (Electrical Desalting Apparatus) by Petrolite. This device and a refinement called the HTHPEDA (High-Temperature, High-Pressure Electrical Desalting Apparatus), are used to approximate the refinery desalting process. Both of these devices are patented, and we have no permission to reproduce their general design here. However, the principle of their operations is based on the operation of refinery desalting units, which can be illustrated (see Fig. 3–9).

Chemicals are screened in these units using a procedure that attempts to reproduce the conditions existing in the refinery unit. In the refinery the temperature of the desalting unit is maintained at approximately 190° F, and a potential is maintained between the two electrical grids of ≈ 25,000 volts A/C. Clean water is fed into the desalting unit along with the crude oil feed in a ratio of 5% water to 95% oil, volume to volume. The crude passes through the grid along with the wash water and is subjected to the alternating current. This current (combined with temperature) favors the breaking of emulsions formed in the water/oil mix, and washes the salts from the oil. The effectiveness of the procedure is determined by the grind-out method mentioned above.

Field Blending Practices

Field blending practices are still performed by specialty chemical company technicians. Normally these individuals bring a standard set of product line chemicals and intermediate chemicals that can be blended to produce the desired treatment in field applications. If the standard set of products, upon screening, is shown to be less than optimal for the application, blends are made from the intermediate chemicals. The blends produced in this fashion are added to the product line if they represent a dramatic improvement over the existing products. Thus, product proliferation is a problem faced by all specialty chemical companies. However, this methodology has been very

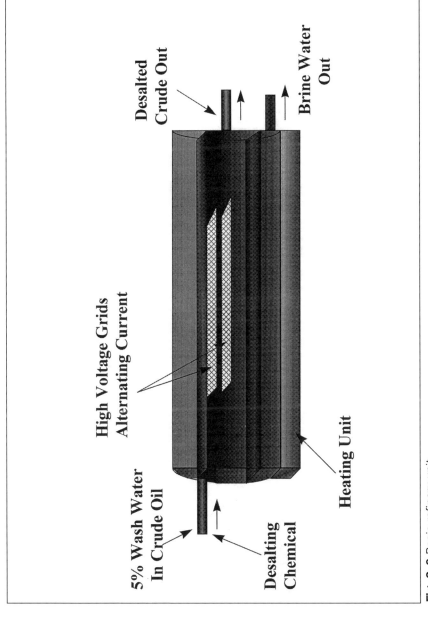

Fig. 3–9 Basic refinery unit

successful in numerous applications, and often represents a substantial chemical treatment cost savings to the oil producer.

Summary

The preceding chapter discusses the typical test methods employed by chemical treating companies for the classification and treatment of oil field emulsions. It is not intended to be an exhaustive listing of the methods used, but rather a description of typical testing that has been accepted by most as standard. Most deviations from these standard methods are small, and include using equipment like automated Karl Fisher water titration of the treated oil. The fact that these methods are used by such a wide variety of companies is a testimonial to their effectiveness.

Problems

3–1. Give three structures that might be commonly found in crude oils, and give each a name based on the structural type.

3–2. When polar functional groups interact with cations or anions in the water phase, what products of water are found?

3–3. How can you explain the existence of two dodecahedron forms for hydrogen bonded water?

3–4. What criteria must be satisfied in order for London forces of attraction to take place?

3–5. Rationalize why the stronger forces involved in ionic and hydrogen bonding can sometimes be overshadowed by inductive interactions.

3–6. Write a cookbook procedure for a grind-out test on a water-in-oil emulsion.

3–7. What type of solvent is used for a grind-out procedure, and what observations are used to determine the effectiveness of the chemical treatment?

3–8. What visual observations are made during an oil in water de-emulsification test, and what method or methods could be used to remove some of the subjectivity of the test procedure?

3–9. Give the meaning of the acronyms HTEDA and HTHPEDA. What field application are they intended to approximate?

References

Barrow, Gordon M. *Physical Chemistry.* 2nd ed. New York: McGraw-Hill Book Co., 1966.

4
Oil Emulsion Breakers

Water in Oil

The types of chemistry involved in emulsion breaker technology represents a very broad range of structural analogs. In this chapter some representative examples of these structures will be presented, and some of their structural activity relationships will be discussed. Some of the early work in this technology was centered around calcium salts of fatty acids, which were derived from saponified products of animal fats (see Fig. 4–1).

These fatty acid salts were heat dried until they formed powders, and sold as laundry powders. It is interesting to note that these products also had been home-made for many years and were called lye soap. From the discussion of the types of chemistries responsible for the formation of emulsions in oils, it would appear that the bipolar chemistries involved in their formation could also be useful in their resolution.

In order to understand why these chemicals are effective as emulsion breakers, it is necessary to examine what happens when they are added to water-in-oil emulsion systems. An interesting property of these fatty acid calcium salts is that they are easily dispersible in oil systems. Additionally, although they have been dried, coordination water is still present in these laundry powders. Thus when they are added to oils, they tend to disperse as micelles (suspensions) with a highly concentrated internal phase of calcium ions surrounded by radially projected nonpolar fatty tails. These micelles act as additional containers for water that normally diffuses between emulsion aggregates.

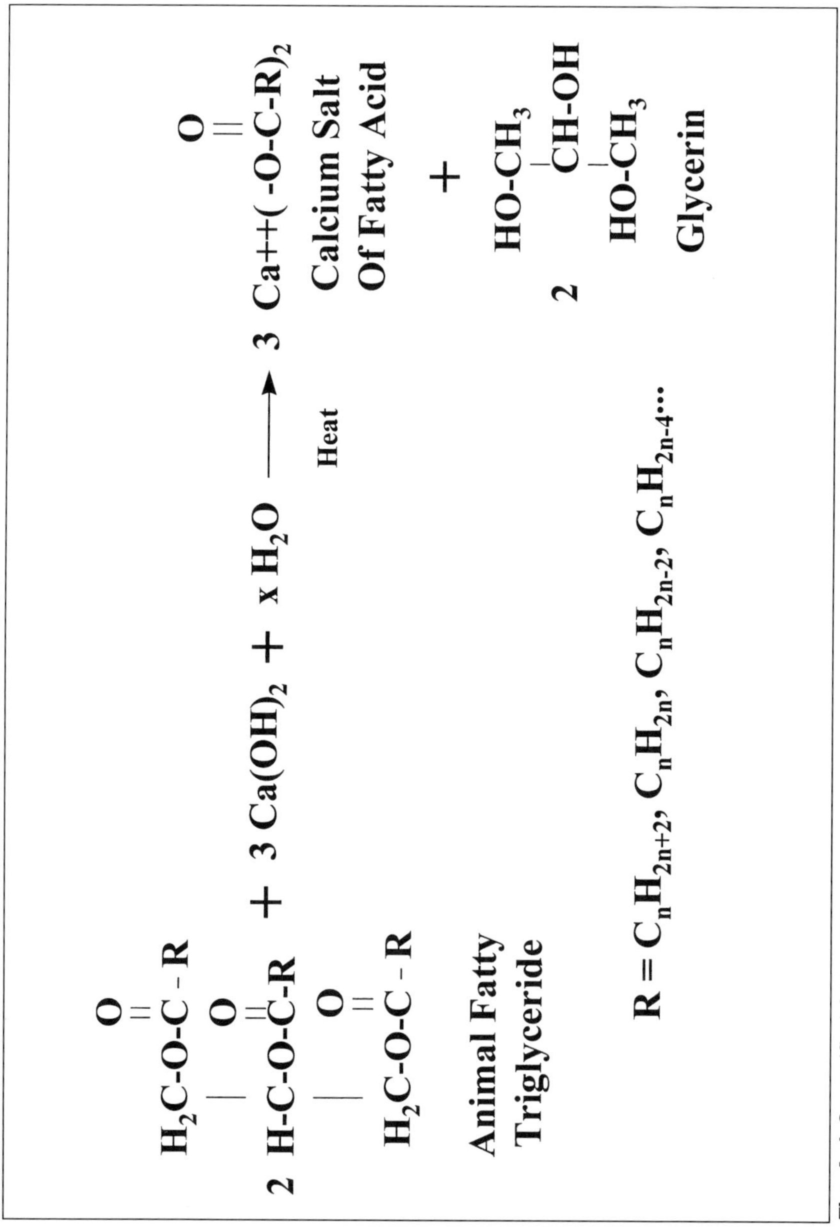

$$
2\ \begin{array}{l} \text{H}_2\text{C-O-}\overset{\displaystyle O}{\overset{\|}{\text{C}}}\text{- R} \\ \text{H-C-O-}\overset{\displaystyle O}{\overset{\|}{\text{C}}}\text{-R} \\ \text{H}_2\text{C-O-}\overset{\displaystyle O}{\overset{\|}{\text{C}}}\text{- R} \end{array} + 3\ \text{Ca(OH)}_2 + x\ \text{H}_2\text{O} \xrightarrow{\text{Heat}} 3\ \text{Ca++}(\text{-O-}\overset{\displaystyle O}{\overset{\|}{\text{C}}}\text{-R})_2
$$

Animal Fatty Triglyceride

Calcium Salt Of Fatty Acid

$+$

$$
2\ \begin{array}{l}\text{HO-CH}_3 \\ |\\ \text{CH-OH}\\ |\\ \text{HO-CH}_3\end{array}
$$

Glycerin

$$R = C_nH_{2n+2},\ C_nH_{2n},\ C_nH_{2n-2},\ C_nH_{2n-4}\cdots$$

Fig. 4-1 Saponified fatty acids

The resulting increase in equilibrium diffusion increases the probability that collisions between unprotected water vapor will produce unstable water droplets. Thus, the diffusion process that acts to maintain internal phase concentrations is accelerated by the addition of a micelle of higher internal ionic concentration. This mechanism is not generally agreed upon by those involved in research related to emulsion breaker technology. A view that is often stated, and held to be true by many researchers, is that the added chemical actually penetrates the emulsion, destabilizing its surface, which allows it to coalesce with a similarly destabilized surface. However, in the case of calcium salts of fatty acids, there appears to be no reason for this to occur, since the interchange of fatty acids at the emulsion interface would not likely destabilize the surface. Figure 4–2 illustrates the two options.

Later developments in water in oil de-emulsification technology built on this chemistry, and acid sulfates and their salts were added to the list of chemical structures used for this purpose. This class of chemistry provides a stronger anionic functionality than the fatty acids, and with the appropriate alkyl substituents groups present they provide for better dispersion in the oil phase. According to C.R. Noller, *Textbook of Organic Chemistry*, "the earliest of the semisynthetic wetting agents were the *sulfated fats and oils* (formerly called sulfonated), which have been in commercial use for over one hundred years. When an unsaturated fat is treated with sulfuric acid, the double bonds add sulfuric acid to give the hydrogen sulfate of the hydroxy acid." Figure 4–3 shows the product of the reaction of sulfuric acid and oleic acid fraction of a triglyceride and the product of neutralization with sodium hydroxide.

Regardless of the strength of the anion, there still appears to be no reason to expect that an exchange of the bipolar groups at the interface of the emulsion will cause it to be destabilized. On the contrary, if the ionic interaction is strengthened by this exchange, then the emulsion should become more stable. Therefore, the diffusion mechanism appears to be the most satisfactory description of the process leading to emulsion resolution.

Nonionic Surfactants
Polyether products

After World War II, a whole branch of chemistry was opened to companies involved in surfactant (surface-active agents) technology. With the advent of condensed polyether made possible by large-scale production of ethylene and propylene oxides, a new class of nonionic detergents began to appear. The condensation products of the ethylene oxide were found to be water soluble (more appropriately nearly molecularly dispersible), and the high reactivity of the oxirane ring made it very useful in a host of chemical reactions. It was found that the propylene oxide gave poly condensation products that tended to be oil soluble (nearly molecularly dispersible).

Combinations of the ethylene oxide and propylene oxide were made in every conceivable ratio. Thus, this new chemistry proved so versatile that solubility characteristics could literally be designed into a particular product.

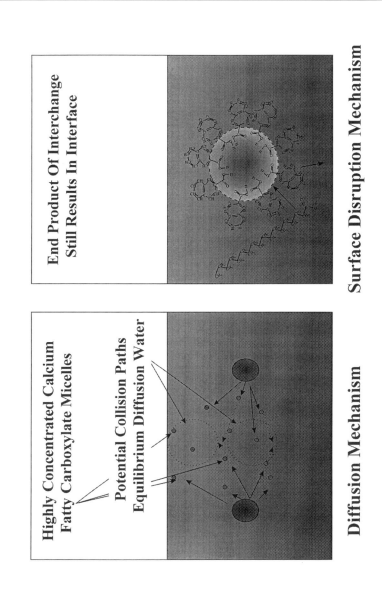

Fig. 4-2 Possible micelle mechanisms

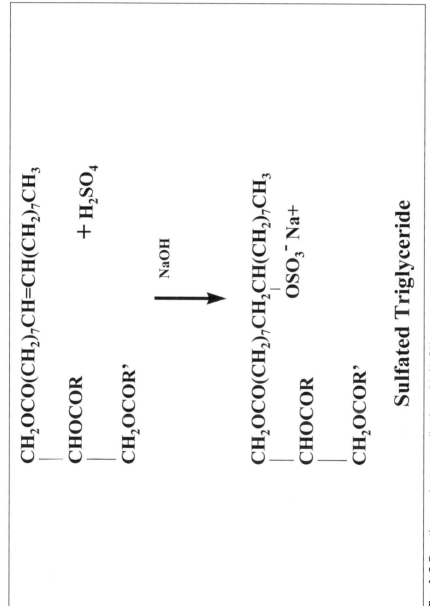

Fig. 4–3 Reaction product neutralization with NaOH

About the same time chlorinated propylene oxide (epichlorohydrin) became available, which allowed the production of several other oxirane-containing compounds.

One compound proved to be particularly useful in producing multi-functional derivatives of propylene and ethylene oxide condensation products. This product is called bis-phenyl-diepoxide. This bifunctional derivative was found to be extremely useful for tying together combination products of ethylene oxide and propylene oxide to form extremely powerful de-emulsifier chemicals. Figure 4–4 presents a listing of some of the structures of epoxides used in the manufacturing of de-emulsifier products used by specialty chemical companies.

The three-member oxirane ring is highly strained, and consequently opens readily when attacked by acid or base. The main difference between the reaction products derived with acid catalyst versus base catalyst is the cleavage site between oxygen and carbon of the ring. Base catalysis occurs at the site of the least substituted carbon atom in the ring (e.g., carbon with fewest alkyl substitution), while acid catalysis occurs by proton attack of the oxirane oxygen. The types of functional groups that epoxides can react with are many and varied, including those groups that can form metal anion salts, carboxylates, phenolates, alkoxides, amines, and amides, to name a few.

Generally the reaction is carried out using an activated starting molecule; potassium salts of dehydrated alcohols are commonly used. The alcohols are mixed with potassium hydroxide, heated to remove water of reaction, and the epoxide is then fed into a stirring reactor under pressure. As the reaction proceeds, the pressure drops, indicating that the oxide has reacted. A typical reaction product of the alkoxide procedure is shown in Figure 4–5.

Figure 4–5 shows how the propylene oxide is opened by a base-catalyzed reaction. The majority of reactions used by specialty chemical companies for the manufacture of nonionic surfactants involve the base catalysis method. The product of this reaction tends to be oil soluble and water insoluble. The condensation product of ethylene oxide, as illustrated in Figure 4–6, will produce a repeating polyethyl ether $(CH_2CH_2O)_n$, that exhibits water solubility (molecular dispersibility).

Note that the reaction of ethylene or propylene oxide take place by successive addition of molecules of oxide; this is generally referred to as condensation reaction. When the supply of oxide is depleted, the reaction is terminated. Once the pressure has dropped following the addition of propylene oxide, a new charge of oxide can be added and the condensation resumes. Block condensation products can be synthesized in this fashion. The differing solubilities of the polyoxide products can therefore be adjusted by alternating charges of the two oxides that produce blocks of propylene and ethylene polyethers. This provides a very powerful synthetic tool for producing a host of chemicals with progressively variable and controllable solubility behaviors.

Figure 4–7 shows an example of a block condensation polyether made from ethylene and propylene oxide. It should also be noted that mixed oxide

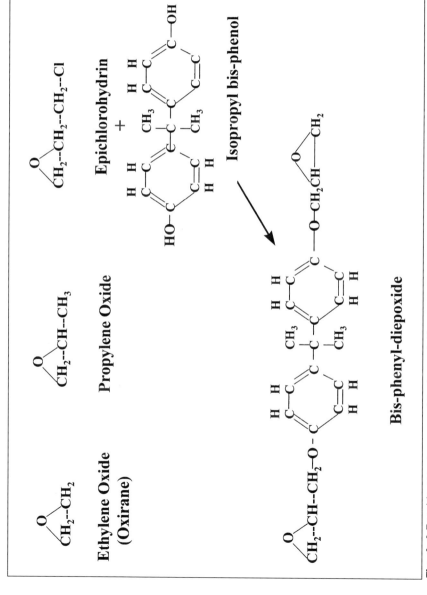

Fig. 4–4 Epoxide structures used in de-emulsifiers

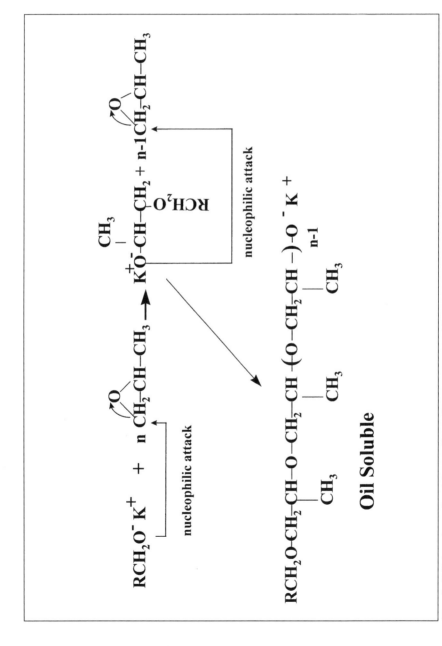

Fig. 4–5 Potassium alkoxide reaction with propylene oxide

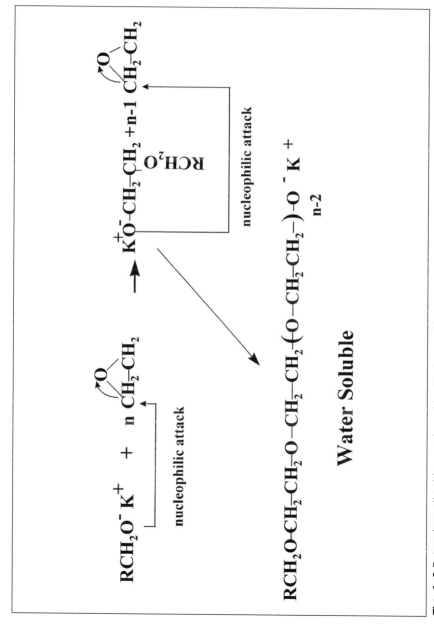

Fig. 4–6 Potassium alkoxide reaction with elthylene oxide

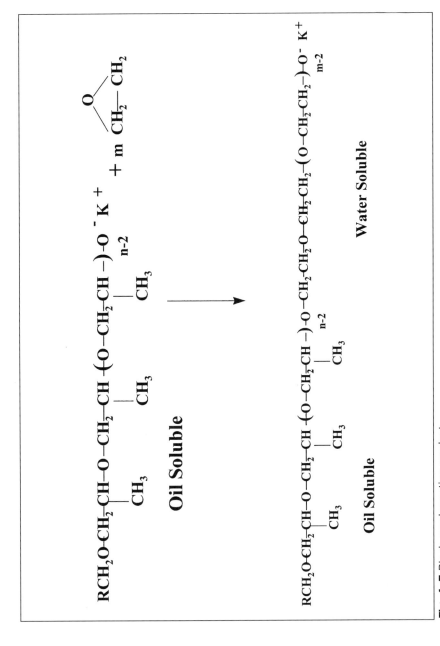

Fig. 4-7 Block condensation product

products can be produced by charging both ethylene and propylene oxide to a vessel containing an activated nucleophile (alkoxide, carboxylate, phenolate, etc.). The result of this procedure is a random collection of polyethers whose solubility characteristics are determined by the reaction rates and charge ratios of the two oxides.

Nucleophiles

Thus far the discussion has been confined to the polyether condensation products of ethylene oxide and propylene oxide, and not much has been discussed about the nucleophiles or parent molecules. Thus in the following discussion the starting molecule that initiates the poly condensation of the oxides will be referred to as the parent molecule.

Early in the development of the polyether condensation products, extrapolations from the activity versus structure of hydrolysis products of fats and oils were integrated into synthetic starting points for oxide additions. This proved to be a very valuable area of synthetic utility. Continued efforts were made to produce chemicals that provide multiple addition sites for oxide addition. One of the early products derived from this effort was the polyether condensation product of formaldehyde-condensed alkyl-substituted phenols. A representation of this product is shown in Figure 4–8.

The polyether adducts of the condensed alkyl phenols were found to be very active as a de-emulsifiers, and the natural progression of block polymers were also made. The series of products made from the condensed alkyl phenol resins were found to affect a very rapid water drop when added to water-in-oil emulsions. Thus the multiple reactive sites of these resins appeared to be key to the activity of the deemulsifier, and a large variety of multifunctional parent molecules was employed as starting materials.

Meanwhile, research had indicated that the molecular weights of the polyethers and block polyethers played an important role in the effectiveness of the de-emulsifier in the process of resolving very small emulsions. This is where the diepoxides appeared to provide a synthetic avenue, and because they could be added at various points in the process, further reaction with the oxides could be continued. Figure 4–9 shows a typical procedure for the formation of such a product. It should also be noted that once the polyether reacts with the diepoxide, a secondary ($CHR_2O\text{-}K^+$) alkoxide becomes available for further condensation with ethylene or propylene oxide. The secondary alkoxide is less reactive than a primary ($CH_2R\text{-}O\text{-}K^+$), but catalytic amounts of acid make it more reactive.

The products made from the diepoxide reaction, and subsequent reaction products of additional ethylene and propylene oxide, continued to show improved emulsion resolution properties. Additional products were made from multifunctional parent molecules, and several good products have been made using this technique. Thus the research still continues, and parent molecule oxide reactions continue in the quest to produce the ultimate product.

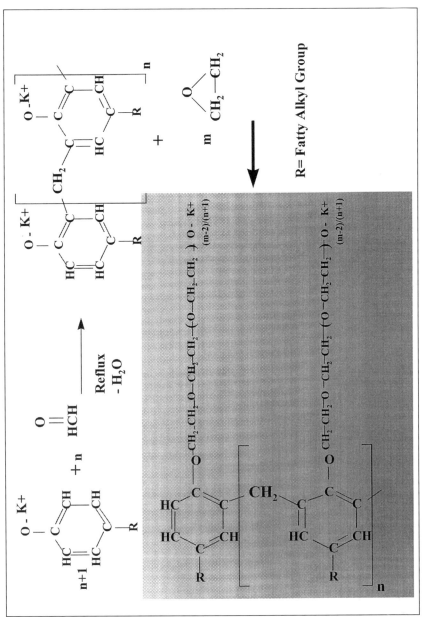

Fig. 4–8 Polyether condensation product of formaldehyde-condensed alkyl-substituted phenols

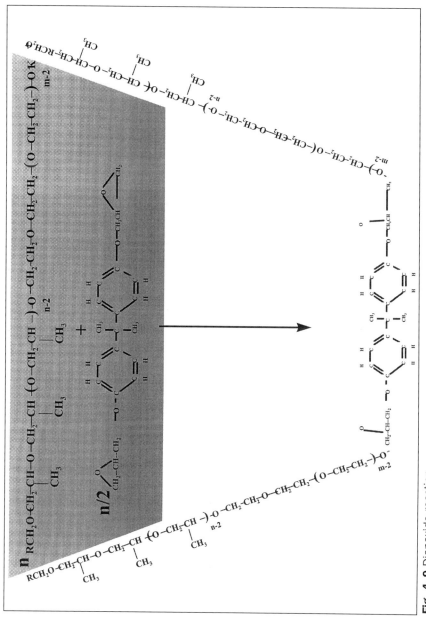

Fig. 4–9 Diepoxide reaction

Because these products are so numerous, it is best to make note of the representative features illustrated in the above examples. However, two very important points should be reemphasized from this discussion:

- Multifunctional polyethers provide rapid water drop
- High molecular weight polyethers exhibit the ability to resolve very small (more stable) emulsions

Summary

The preceding chapter gave an overview of some of the chemistry involved in the development of products intended for the resolution of water-in-oil emulsions. This discussion is intended to illustrate the general classes of products used for emulsion resolution rather than an exhaustive listing. The evolution of the chemistry from the early basic hydrolysis products of fatty triglycerides to the complex mixtures of condensed polyethers and multi-functional parent molecules was discussed in terms of its effect on the properties of the end product de-emulsifier. An emphasis on the diffusion-driven mechanism of demulsification was offered as an alternative to the more commonly held view of emulsion penetration and surface tension reduction. Finally, this discussion also set out to illustrate the extremely flexible nature of the condensation reactions of the epoxides.

Problems

4–1. Give the structure for a fatty triglyceride, and the products derived from its treatment with a strong base in the presence of water.

4–2. Discuss the merits of the surface penetration mechanism for emulsion disruption, and give reasons why, once the de-emulsifier reaches the interface, the emulsion becomes unstable.

4–3. Discuss the reason a bipolar molecule possessing many of the same characteristics as the bipolar molecule stabilizing the interface would be expected to disturb the emulsion interface.

4–4. Discuss the reasons for the relatively high polarity of polyethylene oxide, and the lower polarity associated with polypropylene oxide.

4—5. What differences in solubility will be observed between a randomly reacted mixture of propylene and ethylene oxide versus a block copolymer produced by sequential reactions?

4–6. What are the effects on de-emulsifier activity conferred by the addition of epoxides to a multifunctional parent?

4–7. In what ways does the molecular weight of the poly-epoxidized product affect the treatment effects of de-emulsifiers?

References

Mysels, Karol. J. *Introduction to Colloid Chemistry.* 1st ed. New York: Interscience Publishers, Inc., 1959.

Noller, Carl R. *Textbook of Organic Chemistry.* 3d. ed. London: W.B. Saunders Co., 1966.

5
Water Emulsion Breakers

Oil in Water

Oil-in-water emulsions are representative of systems that accommodate charge interaction through the external phase. Because water can solvate a considerable concentration and variety of salts, the external phase of an oil-in-water emulsion's conductivity can be greatly altered by the addition of such salts. The change in conductivity of the external phase brought on by the addition of a salt is often followed by a destabilization of the oil emulsions present. Further, the addition of multiple valance metallic salts tends to increase the emulsion destablization effectiveness. Thus zinc chloride, ferric chloride, stannic chloride, and aluminum chloride have been used effectively for emulsion resolution (the sulfates are also effective).

One major physical drawback to the use of these salts concerns their tendency to form hydroxyl by-products. These hydroxyl by-products appear as voluminous gels, which are intimately associated with the resolved oil phase. This situation would not be too bad were it not for the fact that the metallic salts are toxic to both the environment and to catalytic processes used in subsequent refining operations. Thus the metallic salts, although effective, are generally considered unacceptable for use as emulsion breakers in crude oil-in-water emulsions. Consequently, the use of organic salts representing low toxicity to both environment and catalytic refinery processes comprises the primary area of product development.

Organic Polysalt Emulsion Breakers

Early in the development of organic polysalts it was realized that the ideal products would be those that partitioned into the oil phase of the emulsion and did not present a problem to subsequent refining processes. Only a few chemicals capable of such behavior were available at the time this work took place, and of these, the polyamines appeared to be the most suitable. As the work progressed, it was noticed that the charge density (charge to mass ratio) of the product was extremely important to the effectiveness of the product. Thus attempts to incorporate the maximum number of amines into the molecule became a primary focus.

With the advent of epoxide technology, chemically altered and functional amines became available to the specialty chemical industry. The ethanolamines and propanolamines (resulting from the addition of ethylene and propylene oxide to ammonia) and substituted amines came on the market. Once these products became commercially available, a great deal of research activity ensued.

Idealized Organic Polysalts

From the above discussion, it appears that there would exist a structure that would present some maximal charge to mass ratio and possess the appropriate solubility to partition into the oil phase. An early product deserving of particular notice was a condensation product of triethanolamine. This product was the result of a high-temperature reaction that involved the removal of water from a pair of hydroxyethyl groups of triethanolamine to form an ether. Figure 5–1 shows this reaction and some derivative salts.

The condensation reaction to form the poly-triethanolamine has quite frequently been carried to a point that the product is completely gelled, which results in an unusable product. An interesting characteristic of this product is that the closer it is to the gel point, the more active the product is as an oil-in-water emulsion breaker. This phenomenon then suggests that solubility, or more appropriately, dispersibility, is a factor in the treating characteristics of oil-in-water emulsion breakers. At this point it is prudent to restate two factors that seem to be essential for the products effectiveness:

1. High charge to mass ratio
2. Borderline external phase solubility (or dispersibility)

Research continued and a large number of polyamine salts and polyquaternary amine products were produced. Figure 5–2 shows some products that possess to some degree the characteristics mentioned above.

The products depicted in Figure 5–2 have been made, and many are commercially available. They are the result of reactions of epichlorohydrin with polyamines, poly-epichlorohydrin with tertiary amines, di-chloro-alkyls, and many other multifunctional molecules.

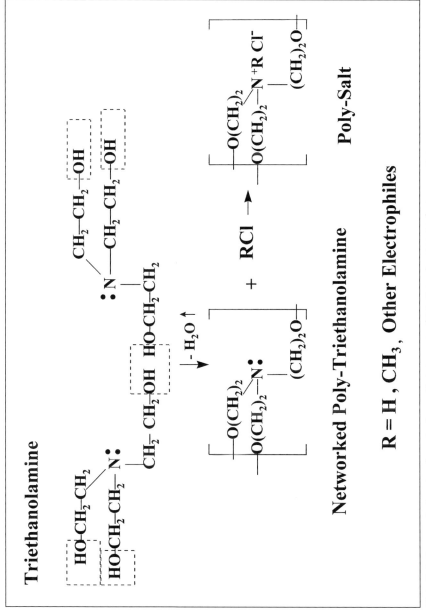

Fig. 5-1 Removal of water from hydroxyethyl groups of triethanolamine to form an ether

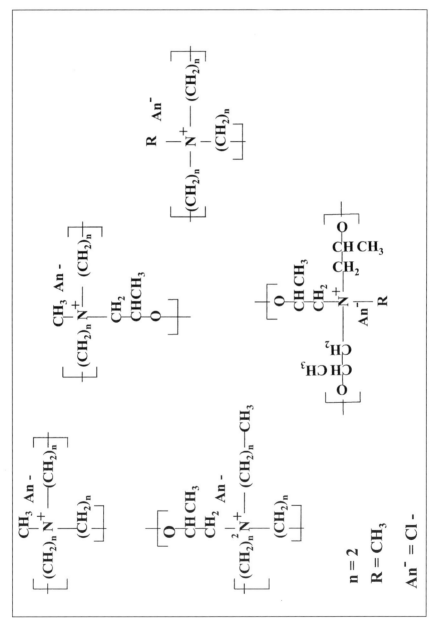

Fig. 5-2 Some idealized oil in water emulsion breakers

Flocculation versus Emulsion Resolution

The preceding discussions pointed out that the effectiveness of oil-in-water emulsions depended upon at least two factors: charge to mass ratio and dispersibility in the aqueous phase. A point that has been alluded to but was not emphasized is that the linear bi-functional polyamine salts are much less active as emulsion resolution chemicals. However when high-molecular-weight, linear polyamine salts are added to resolved emulsions, the oil phase becomes coagulated (flocculated). This flocculation makes the oil that has been recovered from the emulsion easier to handle (e.g., skimming and flotation operations are improved). Thus, products have been synthesized to take advantage of this property.

A host of chemicals have been employed in the synthesis of these high-molecular-weight polyamine salts, including: amino methylated poly acrylamide, poly di-methyl amino propyl methacrylamide, poly di-methyl amino ethyl acrylate, poly ethylene imine, etc. The general class of products consists of linear polyamine salts of high molecular weight molecules, and the higher the molecular weight, the more effective the flocculation process appears to be.

Because of this molecular weight dependence, a variety of methods have been employed to make the highest possible molecular weight products. One method involves the polymerization of the reactive monomers in the internal phase of an oil external emulsion, which in accordance with theory favors high molecular weight polymers. These products are used mainly for flotation aids, and comprise a very large segment of the specialty chemical market. Figure 5–3 shows some of the products comprising this class of chemistry.

Although Figure 5–3 does not specifically show the copolymers, acrylamide, methacrylamide, and the alkyl ammonium substituted copolymers of these monomers are also made. Because of the numerous possibilities for copolymer structures, only the homo-polymer segments are shown. Although several structures have been illustrated, it is important to note that there are countless combinations of starting monomers that can be used for the purpose of making co-polymers, ternary-polymers, tetra-polymers, and higher given the variety of starting materials. Thus Figure 5–4 shows some of the free radically polymerizable monomers available for this purpose.

This discussion should serve as an overview of the class of chemistry involved in the development of products that are useful as de-emulsifiers, flocculants, and flotation aids. An exhaustive listing of the starting materials and the end products is neither practical nor instructional.

Water-in-Oil Emulsion Breakers

Chapter 4 provided a background for the development of water-in-oil emulsion breakers, and two important conclusions were derived from this discussion:

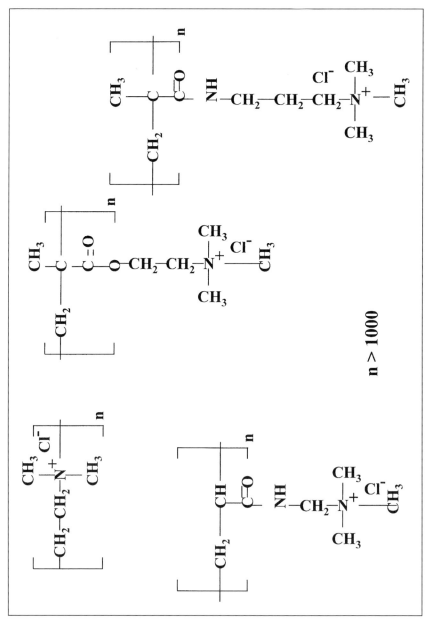

Fig. 5-3 Some representative linear, high-molecular-weight polyamine salts

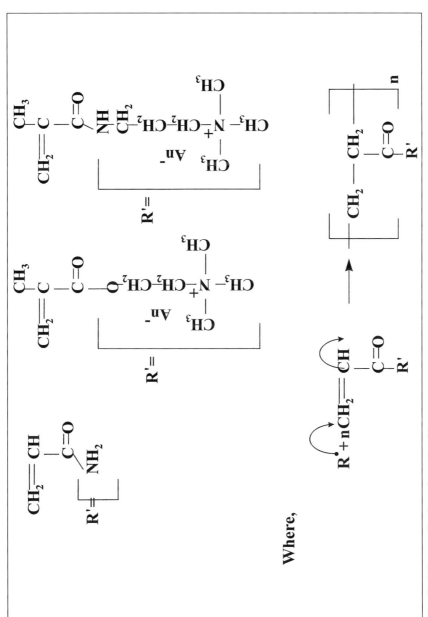

Fig. 5–4 Free radically polymerizeable monomers

- Multifunctional polyethers provide rapid water drop
- High molecular weight polyethers exhibit the ability to resolve very small (more stable) emulsions

In the following discussion the mechanistic reasons for this behavior will be examined, and the focus will be on the nonionic surfactants (polyethers), since they make up the major class of chemicals used for water-in-oil emulsion resolution. In chapter 1 a structural representation of a water-in-oil emulsion formed from naphthoic acid was discussed. This system will be employed in the following discussion to develop the structural relationship of a polyethers activity as a de-emulsifier. The answer to four questions will serve as guides to this development:

1. Is the resolution of a water-in-oil emulsion the result of bi-layer penetration and surface tension disruption, or is it due to the creation of an increased diffusion gradient?
2. Are cationic species within the emulsion extracted by the helical arrangement of the polyethers?
3. Why do polyethers derived from multifunctional parent molecules give rise to rapid water drop?
4. Why do high molecular weight polyethers drop water slowly but possess the power to resolve very tight (stable) emulsions?

The answers to these questions are perhaps best explained by graphical arguments. Figure 5–5 illustrates some of the principles involved.

Figure 5–5 illustrates two possible interfacial orientations of the alkyl-substituted resin polyether after its introduction to a water in oil naphthoic emulsion. The figure on the left requires that the water dispersible polyether portion of the de-emulsifier overcome the cohesive forces of the naphthoic acid to penetrate the water phase. The figure on the right illustrates a more likely arrangement, since the aromatic interactions of the resin with those of naphthoic would require less energy to overcome repulsive forces arising from group interactions. The end result of the second arrangement (the one on the right of Figure 5–5) might be displacement of naphthoic by the resin, aggregation and formation of a second and third layer by the resin, or most likely both results.

In any event, there seems to be no compelling reason to expect that either one or both resulting arrangements would destabilize the emulsion; on the contrary, it would appear that the arrangement on the right would act to stabilize the emulsion. Question 1 above appears to have been answered in favor of the diffusion mechanism, by exclusion of penetration as a viable mechanism. Figure 5–6 illustrates the concept of a diffusion-facilitated arrangement (note that this mechanism has been schematically represented in Figure 3–3).

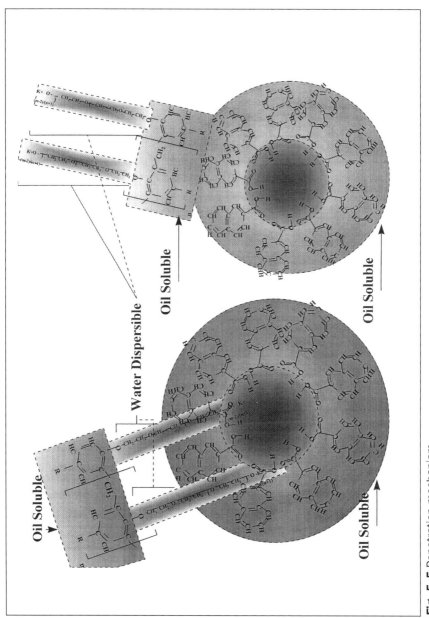

Fig. 5–5 Penetration mechanism

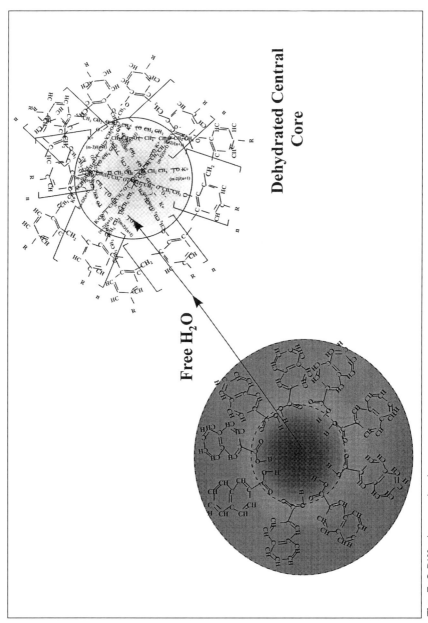

Fig. 5–6 Diffusion mechanism

In Figure 5–6 the added de-emulsifier aggregates itself with the polar polyethylene oxide functions in the interior of the micelle and the fatty sub-stituted aromatic resin radially projected outward into the oil phase. This arrangement provides a potential reservoir for the accommodation of diffus-ing free water, and in fact provides a differential potential that acts to drive the diffusion process. The increase in molecular diffusion of free water increases the incidence of effective free-water collisions on their way out of and toward the drier miceller de-emulsifier reservoir.

Question 2 is also approached graphically; however, the results may be more difficult to visualize. Figure 5–7 illustrates the possibility for coiling of the polyether adduct of the alkyl phenol resin. Various crown ethers exhibit very potent metallic coordination abilities for hexa-coordinate metal cations. When the lone pair of electrons on oxygen are arranged in the fashion sug-gested in Figure 5–7, a pseudo 12-6 (12 carbons and 6 oxygens) crown ether configuration is obtained. This poly ether coil then is capable of coordinating metal cations so tightly that they can be carried into the organic phase of the emulsion.

This still does not answer question 2, but it does suggest the possibility of the pseudo crown ether formation. Further, if the de-emulsifier contacts the external boundary of the emulsion as depicted on the right side of Figure 5–5, a potential driving force for cationic migration is possible. It is possible if it is accepted that the phase boundaries population composition of polar to nonpolar species becomes blurred in the interface interval. Thus, it is likely that cations are exchanged, and that the poly ethers facilitate ionic interchange.

The answer to questions 3 and 4 can be approached by suggesting that miceller aggregates are composed of accumulations of like dipolar species, and there is a thermodynamically favored maximum number of dipolar molecules per micelle. In the case of a linear polyether, this favored maxi-mum provides a numerically greater number of diffusion receptacles than a polyether derived from multifunctional parent molecules.

Although the number of micelles is greater in the linear polyether, the potential diffusion gradient between the emulsion and the micelle is less than one produced by a greater number of polar groups at the micelle's center. Thus, two variables can be adjusted to achieve a balance between the number of micelles and the potential diffusion gradient between the emulsion and micelle. One variable is the molecular weight of the polar/nonpolar polyether, and the other is the number of polar/nonpolar polyethers in a single molecule. Figure 5–8 is an illustration of these two variables, and it also suggests the rela-tionship between diffusion potential and molecular configuration.

Thus plausible answers to all four questions have been given, and a set of general statements can be made that will summarize these answers. Water-in-oil emulsions are resolved by de-emulsifiers through the provision of micelles, which develop potential diffusion gradients between the emulsion and micelle. The larger the dehydrated center of the micelle, the greater is

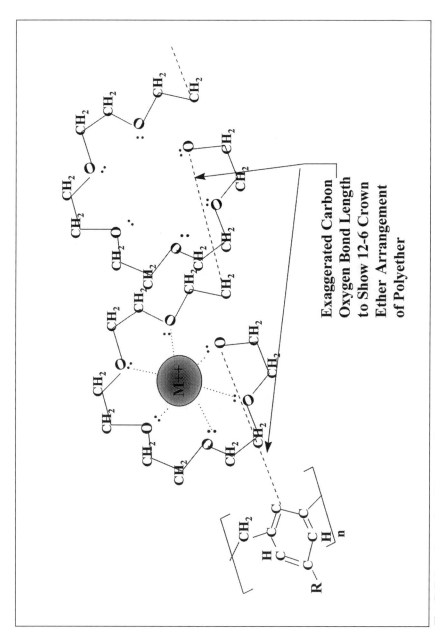

Fig. 5–7 Coordinated metal in polyether cage

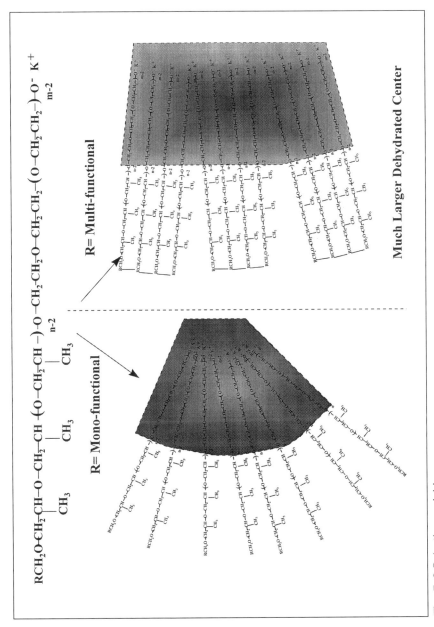

Fig. 5-8 Polyether variables

the diffusion gradient and free-water diffusion rate. Finally, the greater the micelle number, the more likely diffusing free water will reach a dehydrated micelle; conversely, the larger and fewer the micelles there are, the more likely diffusing free water will encounter another diffusing free water.

Oil-in-Water Emulsion Breakers

Oil-in-water de-emulsification is much more completely understood than water-in-oil de-emulsification. Therefore, this section will present an overview of the prevailing theory. Oil internal emulsions present an anionic surface to the external water phase, and the water phase provides cations to maintain charge neutrality.

A good illustration of this stabilized arrangement is depicted by the stearic acid system. Figure 5–9 shows an idealized representation of a charge-stabilized stearic emulsion, charge balanced by commonly occurring calcium cations. Only one layer of the interface is shown for visual clarity, but it should be remembered that there is usually more than one layer of bipolar stabilizing molecules at the interface. It is also important to note that the charged species are in the external or water phase, and are available for interaction with anions introduced into the system.

When metal salts like ferric chloride are introduced, as mentioned earlier, some of the bipolar stearate anion are protonated. Meanwhile, the ferric chloride salts exchange hydroxyl anions for chloride anions in the process of protonating water to form the hydronium cation hydroxyl anion pair.

$$Fe^{+++} + 6\ H_2O \longrightarrow 3\ H_3O+ + Fe(OH)_3$$
$$3\ H_3O^+ + 3\ RCOO^- \longrightarrow 3\ H_2O + 3\ RCOOH$$
$$3\ Ca^{++} + 12\ H_2O \longrightarrow 6\ H_3O+ + 3\ Ca(OH)_2$$
$$3\ Cl^- + 3\ RCOOH \longrightarrow 3\ HCl + 3RCOO^-$$

Note that the end product of the addition of ferric chloride is an anionically charged emulsion surface. Although the fatty carboxylate anion (RCOO⁻) exchanges protons with water, it is in equilibrium, and the emulsion resolves as an oil fatty acid combination. Note also that for every ferric cation added four metal hydroxides result (these hydroxides appear as fluffy flocks). Thus trivalent metallic salt addition to an emulsion system tends to resolve the emulsion with the additional production of metal hydroxide flocks. These flocks are usually hydrated to a high degree and contribute to the problem of disposal.

Polyamine and Quaternium Salts

The polyamine and quaternium salts destabilize the emulsion surface through a mechanism that involves the protonation of the carboxylate anion bipolar surface, but the polyamine or the polyquaternium amine does not form the hydroxyl flock. It is also important to realize that the water is proto-

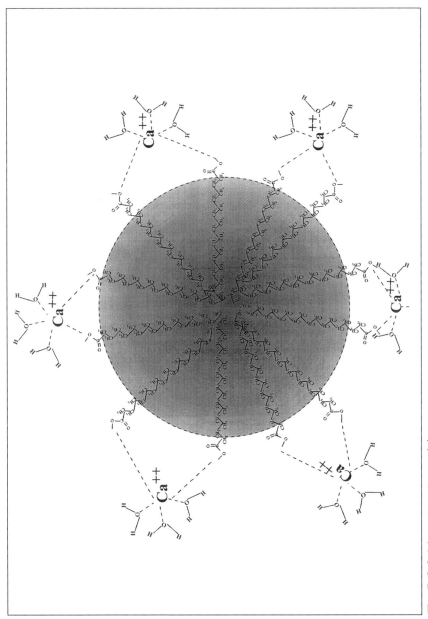

Fig. 5-9 Calcium stearate emulsion

nated to the hydronium ion, as it was the case in the metallic salt process, but that the absence of the trivalent metal cation prevents the formation of flock. The polyamine salts and quaternaries accept the hydroxyl anion and remain soluble in the aqueous phase. In the case of some bivalent metal salts (e.g., zinc chloride) the metal hydroxyl formed is slightly soluble in water, but its emulsion resolution power is less than the trivalent metal salts. Thus, the greater the density of charge residing within a metal salt, polyamine, or polyquaternium salt, the greater is its power to neutralize the emulsion surface.

As mentioned previously, linear polyamine salts and their quaternary analogs are not as effective in breaking emulsions as are the partially cross-linked trifunctional counterparts. This is because the linear molecules tend to extend into the aqueous phase and produce a diffuse charge locus, while the partially cross-linked analogs present a greater charge to solute displacement. The importance of the role water plays in the surface neutralization cannot be overemphasized, nor can the charge to mass ratio or charge to solute displacement of the emulsion breaker.

In chapter 1 it was pointed out that a ratio as little as 1:10,000,000 of molecular concentration of bipolar species to external phase can form an emulsion. Thus, small amounts of emulsifier can produce stable emulsions, and small amounts of cationic de-emulsifier can destabilize these surfaces. Figure 5–10 summarizes much of this discussion.

Summary

This chapter shows the emulsion resolution mechanism of the polyamine salts and quaternaries. No special effort is made to invoke the theory of the Helmholtz layer or to describe the resolution of the emulsion by achieving a charge equivalent to the zeta potential. Although these factors are very much at work in these systems, it is more instructive to show the result of the addition of the metals and poly salts than it is to go into the mathematical rigors of these phenomena.

The main points to remember are that

- the emulsion is destabilized by bipolar protonation
- water must provide the hydronium and hydroxyl species
- emulsion breaker charges must be concentrated at a minimum locus

The description of the mechanisms is not generally adhered to by those involved in this field. The discussion is intended to suggest alternative schools of thought that better fit the observed behavior of the de-emulsification process. The conclusion of this presentation is that diffusion rather than penetration and surface disruption provide a better explanation of the observed phenomena.

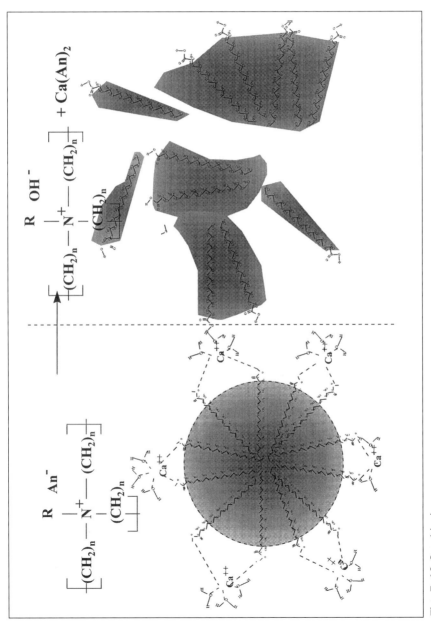

Fig. 5–10 Graphical summary

The results of this view are important enough to be restated from the paragraph above. Water-in-oil emulsions are resolved by de-emulsifiers through the provision of micelles,which develop potential diffusion gradients between the emulsion and micelle. The larger the dehydrated center of the micelle, the greater is the diffusion gradient and free-water diffusion rate. Finally, the greater the micelle number, the more likely diffusing free water is to reach a dehydrated micelle; conversely, the larger and fewer the micelles there are, the more likely diffusing free water is to encounter another diffusing free water.

Problems

5–1. Is the resolution of a water-in-oil emulsion the result of bilayer penetration and surface tension disruption, or is it due to the creation of an increased diffusion gradient?

5–2. Are cationic species within the emulsion extracted by the helical arrangement of the polyethers?

5–3. Why do polyethers derived from multifunctional parent molecules give rise to rapid water drop?

5–4. Why do high molecular weight polyethers drop water slowly but possess the power to resolve very tight (stable) emulsions?

5–5. What result would you expect to see if the diffusing water were accompanied by a portion of the bipolar emulsion phase?

5–6. Suggest an explanation for the spiral configuration of the polyether portions of epoxidized bipolar de-emulsifiers.

5–7. Explain why metal cations would be expected to be coordinated by the spiral polyether configuration.

5–8. Suggest additional structures that might act to coordinate metal cations, and suggest a plausible synthetic route for each.

References

Mysels, Karol. J. *Introduction to Colloid Chemistry*. 1st ed. New York: Interscience Publishers, Inc., 1959.

Section II

Waxes

6
Petroleum Companies and Waxes

Paraffin Wax in Crude Oil

The presence of paraffin waxes in crude oils presents a host of problems to the producer, transporter, and refiner. The build-up of paraffins and asphaltenes represents the organic equivalent of scale formation, and their presence in the formation, tubing, transfer lines, storage vessels, and pipelines can lead to serious problems. The problems associated with their presence range from minor to severe, and depend on their quantity and composition. However, the paraffin waxes are valuable sources of refined products ranging from motor oil to jet fuel.

If one examines the nature of a production reservoir, it becomes evident that these structures behave much the same as a distillation apparatus. When these reservoirs are tapped, the volatile components are often the first to depart their environment. This departure causes a concentration increase of the heavier factions within the confines of the reservoir. Two critical effects result: the pressure drops (and therefore the drive efficiency decreases) and the mobility of the remaining heavy fraction declines. As this occurs, the job of producing the remaining crude becomes increasingly difficult, and special measures must be undertaken to combat these effects.

Additional difficulties arise as the depth of production horizons increase. Because the mobility of lighter crude fractions is considerably greater than the higher molecular weight asphaltene and paraffin wax frag-

ments, the overburden acts as a semipermeable membrane, allowing light fragments upward migration. The overall result is the accumulation of increasingly heavy crude fractions occupying the reservoir.

Paraffin Wax Production Problems

The increased concentration of heavy wax fragments within aging and increasingly deeper reservoirs is just the beginnings of the producer's problems. Variable tubing fluid velocities resulting from viscosity effects, wax deposition in the formation and tubing, and natural choke formation by deposited asphaltenes and waxes all converge to restrict production. Once the crude exits the well, transfer lines, treatment vessels, and storage tanks are all negatively affected by the presence of paraffin waxes intermingled with emulsions, solids, and corrosion by-products. Discontinuous surfaces act as sites for the formation of wax crystals, and solid phase emulsions, combination wax/asphaltene deposits, and gelled crudes frequently appear.

Many of these problems can be effectively resolved by the appropriate application of crystal modifier chemicals; however many factors determine the feasibility of such treatments. As with the application of de-emulsifiers, the placement of these chemicals should be as early in the production stream as possible. Temperature is important to both the process of emulsion resolution and crystal modification and, as we shall see in much greater detail in the chapters that follow, it is one of the primary physical variables determining the effectiveness of chemical treatment.

Paraffin wax deposition within the producing reservoir is an extremely difficult problem to resolve once it begins and, in many instances, it involves the cessation of natural drive production from these reservoirs. Remedial treatments often involve the use of solvent soaks, hot water, or hot oil treatments to revitalize production. Each of these methods are characterized by strengths and weaknesses, but they all involve well shut-in time that is always costly to a producer.

While it is true that wax and asphaltene deposition can be moderated by production rates, it is also true that a trade-off exists between the formation of naturally formed asphaltene and wax chokes. These natural chokes act as speed controls for the production rate of crude oil from a completed well, and if these constraints are ignored, the consequences have a profound effect on the life expectancy of a producing site. If, in the process of bringing crude oil to the surface, the flow velocity is too slow, the probability of wax deposition is increased, since environmental cooling may occur as the fluid passes through low temperature zones. However, if the fluid velocity is very high, the likelihood of asphaltene deposition is increased by the resulting increases in streaming potentials. Thus, the job of the production engineer is complicated by the demands placed on him for maximizing production while maintaining the life expectancy of the wells under his or her supervision.

Organic Deposition Control

Although good production practices can minimize the formation of asphaltene and wax deposits, the economic viability of a well demands a minimum production rate. This production rate may not always be achieved by production engineering techniques, and it is in these cases that chemical treatment becomes a necessity. As mentioned earlier, crystal modifier chemicals are most effectively applied as early in the production stream as possible. The lack of capillary injection facilities complicates the logistics of crystal modifier addition, and often results in the improper addition of chemical under less than optimal conditions.

It is not uncommon to see deposition occurring in the well, and chemicals being added after the wellhead. While this practice may prevent continued deposition in transfer lines and storage vessels, the source of the problem goes untreated. Various attempts to add crystal modifier down the backside are only partially successful, since this intermittent chemical contact with production fluids cannot maintain optimal chemical dose levels.

Removal of Paraffin Wax Deposits

Once wax deposition has occurred, the options available to production and facility engineers involve the use of hot oil, hot water, hot water and surfactant, and solvents. Because usage of these methods has associated intrinsic risks, their application is a matter of considerable concern. The use of hot oil treatments in wax-restricted wells can aggravate the problem in the long run, even though the immediate results appear good. This technique usually involves the heating of stored oil from the production site and circulating it down the tubing to remove the wax deposits.

A closer look at this practice shows that although lighter waxes, which act as mortar for the heavier waxes, are removed, the higher melting fractions become concentrated. This effect is particularly damaging to the near-wellbore area, and repeated treatments may cause severe obstructions that will require increasingly severe treatments to remove.

Hot water and combined hot water and surfactant treatments must be carefully considered prior to implementation, because some producing formations are water sensitive. However, when conditions allow, these procedures are quite effective. These treatments are performed using hot oil trucks that heat the water or water and surfactant and pump it down the tubing to remove the deposits. The hot water technique suffers from the same limitations as the hot oil method, and higher molecular weight waxes (those with higher melting points) are concentrated.

The combined hot water surfactant method allows the suspension of solids by the surfactant's bipolar interaction at the interface between the water and wax. Thus, much of the higher fractions are carried out of the well as suspensions. One additional advantage of the use of combined hot water surfac-

tant treatments over hot oil is that water has a higher specific heat than oil, and therefore it usually arrives at the site of deposition with a higher temperature.

Solvent treatments of wax and asphaltene depositions are very often the most successful remediation method, but they are much more costly. The amount of wax that can be carried by solvents such as xylene or kerosene is a function of the molecular weight of the wax and its concentration. It is not uncommon to find a 10% wax in solvent that is solid at room temperature, so the amount of solvent used can be considerable in some applications. Solvent remediation methods are usually reserved for applications where hot oil or hot water methods have shown little success. Sometimes, combination solvent/surfactant packages show increased performance over solvent alone and are used to enhance the carrying capacity of the solvent. This practice is more common in methods designed for the removal of asphaltenes, but it will probably continue to increase in popularity for wax removal in the future.

Wax and Work-Over

Some locations in the world have such severe problems with wax deposits that wells are shut-in for pulling operations on a weekly or monthly basis. These operations are costly and time-consuming, but production volumes justify them. Very often these procedures could be avoided or reduced in frequency by the placement of a capillary injection string and the continual addition of the appropriate crystal modifier chemical. However, in many of these locations the cost of labor is so low that it makes economic sense to perform the pulling jobs, clean the tubing, and replace the fouled tubing with those refurbished by laborers. The one fallacy in this economic picture involves the cost of lost production as a result of the shut-in time.

In some parts of the world operators maintain redundant transfer facilities to which produced fluids can be redirected while the primaries are disassembled, cleaned, and returned to service. Likewise, the service life of these lines could be extended for long periods by the proper injection of crystal modifiers. Both well-pulling and line-looping procedures present significant risks to the field operator, since the likelihood of a sudden catastrophic pressure increase could result in tubing or line rupture.

Physical and Mechanical Wax Control

Several methods of wax control are practiced by production operations, but the shipment of crude over long distances under the ocean, across mountains, through arctic cold, and across sweltering deserts demands significant planning and forethought. After production fluids exit the wellhead, the problems of maintaining a free-flowing conduit for its transport become paramount. The wax content of some production streams is so high that special methods are employed to keep the fluids liquid. Line heaters represent one of the more effective methods of providing for fluid mobility; however, their design, energy requirements, and operation involve considerable costs.

Although line heaters can be successfully employed from the wellhead to gathering facilities, the physical nature of the crystallizing waxes has not been altered. This lack of alteration can be a problem once the fluids are sent to storage. The prevailing conditions of temperature and fluid movement within the storage battery favor the formation of wax crystals and lead to gels and sludge. Thus the problems of waxes in the production stream have now become the problems of the gathering facility.

Line pigging is also practiced frequently by crude handling operations. This practice requires that launching and capture sites be engineered into the transfer facilitiy's design. Pigging operations are conducted with and without incorporation of solvents and chemicals, and the retrieved material blockages are most often directed to waste streams. The variety and degree of sophistication of pigging devices is staggering, ranging from simple projectiles to devices with onboard telemetry, but a common denominator to each is that they are employed after damage has been detected. Their after-the-fact application then relegates their implementation to a remedial method.

Coiled-tubing technology has become an important means of conducting well clean-up procedures. This technology involves the redirection of well production to fluid collection facilities or flaring operations while the coiled tubing is in the well. Heavy coiled tubing reels are placed at the wellhead by large trucks, the well fluids are diverted, and high-pressure nozzles on the end of the coiled tubing are placed in the well. Tanker trucks filled with solvent provide the high-pressure pumps with fluids that are used to clean the well tubing as the coiled tubing is lowered into the well. The value of this method is apparent in many areas of the world, since certain integrated production companies maintain a fleet of coiled tubing trucks that remain busy a large percentage of the time.

Wax and Crude Oil Transport

Many of the production methods above fail to answer the problem of ownership transfer once the crude oil reaches field storage facilities. Pipeline companies and crude oil movers have a responsibility to protect their equipment from damages that can occur from the purchase of crude oils that are below certain specifications of water and wax content. These specifications usually demand that the crude oils they purchase fall within a range of pour points that will not result in damage to their facilities. This places the burden of the cost of chemical treatment on either the producer or the transporter.

If the transporter is required to purchase chemical prior to shipping the crude, this cost is subtracted from the price he is willing to pay. On the other hand, if the producer takes on the cost of treatment, the selling price he requests will be increased to offset his added cost. Either way chemical treatment programs are usually a necessity. This situation then raises the question of the most economical and effective placement of crystal modifier products in the production or transport stream. Throughout this discussion it

should have become evident that the placement of these chemicals should be as early in the production stream as possible. The mere act of transferring the problem from an upstream location to one downstream is not helpful, and in most cases, it is counterproductive.

Waxy Crude Oils and the Refinery

It is ironic, in many respects, that those crude streams that are most difficult to produce and transport are often the source of much of the profit realized by the refinery. The successful production and transfer of waxy oils to the refinery is essential for the profitability of the refinery. Although storage of these crude streams can be difficult since they produce gels and sludge in massive storage tanks, crude movement facilities generally apportion the refinery influent stream to maintain a good turnover. Thus the waxy feed stocks are blended with lighter crudes prior to shipment to internal refinery processes.

Once in the refinery, waxy crude blends are fed into the desalting units, where they are prepared for the distillation processes. The atmospheric and vacuum distillation processes remove the lighter fuels, and the bottoms are shipped to solvent processes to separate the waxes from the asphaltenes. The wax fractions are then subjected to cold solvent additions that produce wax slurries that are vacuum filtered to remove motor oil. Depending on the demand for motor oil and wax, the atmospheric and vacuum tower bottoms can be diverted to catalytic cracking facilities. Waxy crudes are valuable sources of feed stocks to catalytic crackers, and produce high yields of valuable fuel products.

The preceding discussion gives an overview of some of the problems faced by the production, transport, and refining of waxy crude oils. One of the most important points to emerge from this discussion is the need for detailed planning. Planning by the each facility from the producer to the refiner is necessary to avoid or remove the effects of wax accumulation. The discussion also points to the value waxes have to the end-product mix, and why it is so important to provide adequate measures to afford its availability to the refinery.

Waxes and waxy crude oils present a challenge to those handling them, but they have value-added qualities that emulsion streams generally lack. The following discussions will develop a background for the understanding of the nature of wax problems and provide an insight into how they may be handled.

Treatment Problems

The complex mixture of chemical species present in crude oils makes it necessary to consider several factors when designing a treatment program for field application. Waxes may be the main issue, but compounding the treatment difficulties are the presence of other complex multiple aggregates such as emulsions, asphaltenes, silt, corrosion by-products, and inorganic scales. Most of these materials can be incorporated into the wax, or conversely the waxes can form in them. It is common to see combinations of these problems in a variety of production areas, refinery storage, and crude oil shipments. The challenge is to determine which problem to attack first in order that the most efficient solution may be implemented.

Wax and Emulsions

A combination wax and emulsion is present in a system, and the viscosity of the fluid mix is very high and not reduced by heat and de-emulsifier alone. How can an effective treatment program be implemented to resolve the emulsion and treat the wax? First let's take a look at some likely structural aggregates that might be present in this mix. Solid phase emulsions are known to occur in which the wax has accumulated at the interface between the water and oil. This is not so surprising, since both the molecules that stabilize the interface (e.g., fatty acids) and waxes have alkyl groups that can combine (see Fig. 6–1).

As can be seen from the diagram, the wax molecules can combine with the fatty tails of the carboxylic acids stabilizing the interface. Considering this view and the fact that the emulsion is not resolved with heat and a de-emulsifier, then an obvious approach would be to apply a crystal modifier to interfere with these interactions. So a completely reasonable approach would be to include the crystal modifier as part of the emulsion resolution treatment. Another aspect of this treatment is that a crucial temperature would be expected, since the melting point of the wax should be approached when attempting to de-emulsify this oil.

Solids and Waxes

Solids and waxes frequently combine to form inclusion aggregates, which complicate the treatment of waxes by the addition of a crystal modifier alone. In these cases it is helpful to consider a likely arrangement of this complex aggregate. It helps to look at both options of whether the forming wax crystal serves as a collection point for the solid, or the solid serves as a nucleation point for the onset of wax crystallization (see Fig. 6–2). In either case, temperature is a prime consideration, because the solid remains solid over the typical temperatures prevalent under most field conditions.

It also helps to know the nature of the solid, for example, whether it is an asphaltene, sand, scale, or corrosion by-product. This is usually determined by hot filtering the crude from a 5:1 xylene:oil blend, and analyzing the filtered solid. If the solid is silt (sand) it will have an anionic character because of its SiO_8 unit-cell structure, thus a cationic surfactant such as an amine combined with a fatty sulfate will likely show activity in dispersing the silt.

Once the silt is dispersed, the wax crystal problem becomes primary, and again temperature is crucial to the application. If the temperature is below the crystallization temperature of the problem wax, an approach other than crystal modification will be required to treat the oil. The advantage of using an

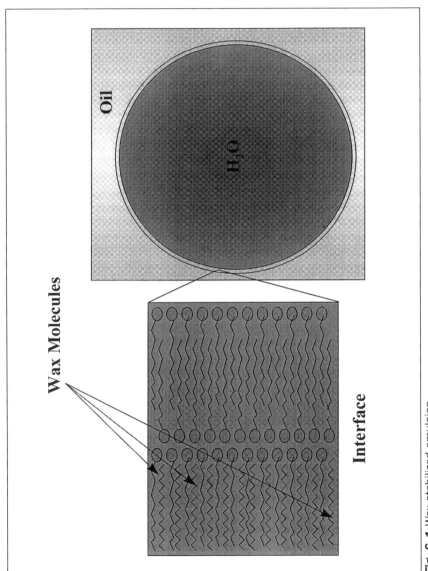

Fig. 6–1 Wax stabilized emulsion

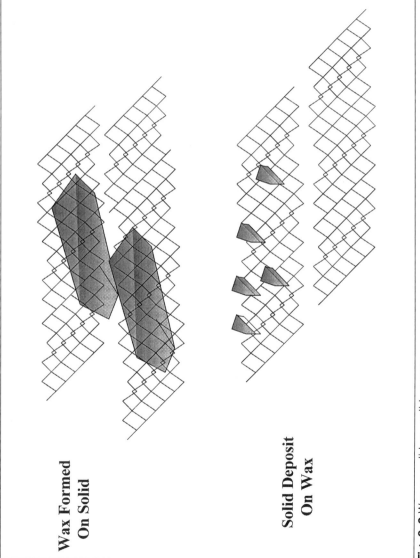

Wax Formed On Solid

Solid Deposit On Wax

Fig. 6–2 Wax on solid vs. solid on wax

amine salt of a fatty sulfate is that it might also act to disperse the wax crystals. If the solid is a corrosion by-product, it is generally in the form of an oxidized metal salt, and again these salts can be dispersed with an amine sulfate.

Waxes and Asphaltenes

Asphaltenes present a special case because they are high-molecular-weight complex hetero-atomic poly-cyclic species that exhibit a polymorphic aggregate behavior. These structures are almost always found in association with waxes when they are retrieved from wells, storage tanks, or pipelines. They tend to form because of the removal of a naturally occurring mixture of resins and maltenes that act as natural dispersants for the asphaltene. The removal of these naturally occurring resins and maltenes can be accomplished by such small shear forces as the off-gas of light-end alkanes (e.g., methane, ethane, butane, pentane, and hexane).

Resins and maltenes can also be destabilized by inductive forces at the pipe surface or inductive interactions with other asphaltenes or paraffins. This may explain why they are so often found associated with paraffin waxes. The nature of the maltenes and resins that stabilize the asphaltene resemble the subunits of the asphaltene and are likely precursors of the asphaltene. An approach to a treatment of asphaltenes would be to approximate the structure of these stabilizing molecules. A possible key to a treatment would then involve the use of materials that have fatty character and a hetero-atomic parent molecule.

Some Wax Treatment Considerations

Frequently the temperature of the system to be treated is adequate for the incorporation of the crystal modifier into the growing wax crystal, but the mixing of the polymer modifier is inadequate. This problem is sometimes overcome by the inclusion of a dispersant to the modifier product. The dispersant divides the polymer into smaller fractions that can mix more readily with the crude under low conditions of shear. Dispersants that are used, however, must be compatible with the chemistry of the crystal modifier. Typical dispersants for these applications are usually nonionic poly-ethers manufactured from ethylene oxide, propylene oxide, or mixtures of both.

Other concerns that face the chemical company include the prevailing temperature of the chemical use point and the physical state of the chemical product at this temperature. Very often the concentrated crystal modifier chemical is solid under prevailing temperatures and must be diluted or otherwise modified to enable its pumping. Physical modifications such as powder or emulsion are among some of the options available, but delivery systems (pumps) would also require altering, and this is not generally acceptable to the customer.

Testing Methodologies

Pour-point testing

Testing procedures for the determination of wax problems in crude oils have evolved over the years from simple visual tests to complex and elaborate methods requiring considerable equipment and expense. One of the early tests employed involves the heating of a crude oil sample in a pour-point tube (usually a thin-walled Pyrex tube with 50 ml capacity) to a temperature of about 150° F for approximately 30 minutes (to assure the waxes have melted). After the 30-minute heating period, the tube is fitted with a thermometer and cork arrangement that allows the thermometer to be immersed in the oil, and the whole arrangement is allowed to cool to room temperature.

The tube containing the sample is turned on its side and held at a 90° angle to the upright for 5 seconds, and observed for fluid movement. This process is repeated in 5° F increments as the temperature drops. After the sample temperature has equilibrated at room temperature, it is placed in a refrigerated system and air cooled to a temperature at which it no longer flows, when held at 90° to the upright for 5 seconds. This temperature is then reported as the pour point of the sample (see Fig. 6–3).

Although this test is still used, and provides a good means of screening chemical treatment chemicals, it provides only the barest information about the nature of the wax problem. Although the oil sample has stopped flowing within the 5-second period at its pour point, this method provides no information about what viscosity this represents. Variances in viscosities for different samples at their pour points can range several thousands of centipoise.

Another shortcoming involves the formation of a wax skin at the liquid air interface. Since the partial pressure differential at the liquid-air interface is substantially different than the partial pressure differentials in the bulk fluid, a wax skin can form at this surface. This wax skin then impedes the flow of the sample when held at a 90° angle to the upright. Thus, the pour point measures impedance to flow under very mild conditions of shear and therefore might be better characterized as a forward yield point temperature. Nevertheless, pour-point testing is a very important method, and it still represents one of the fastest methods of screening crystal modifier chemicals. The main point to keep in mind is that, at best, pour-point testing provides a qualitative technique that is important for initial screening tests.

Deposit testing

Over the years, several testing procedures have been developed for the measurement of wax deposits produced by crude oils. A complete listing of these test methodologies is not possible, since they are often proprietary to the laboratories that employ them. But a brief description of some of the publicly available methods can be given.

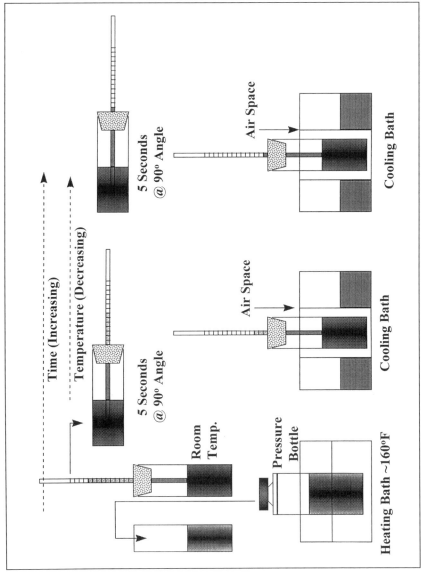

Fig. 6-3 Pour-point testing

Cold-finger testing

Cold-finger tests have been devised to answer some of the questions about the conditions of temperatures at which deposits form. These tests are often performed using equipment designs that range from quite simple to very elaborate. The basic concept of these tests is that a surface (cold finger) is placed in a sample of heated crude oil, and cooling fluid (provided by a thermostated circulating heating and cooling bath) is circulated through the interior of the cold finger.

The oil is gently agitated about the cold finger with a magnetic stirrer while the oil is maintained at a temperature above its cloud point, and deposits form on the cold finger's surface. The amount of deposit collected after the test has been run is a measure of the problems the field will experience with the oil. The duration of the test can range from 3–40 hours and is highly dependent upon the individuals conducting the test.

The extent to which conditions of pressure and temperature are controlled determine the degree of sophistication of the instrumentation. If duplication of the field conditions of pressure are desired, special device-construction considerations must be taken into account. Under normal conditions of pressure (1 atm) no special construction considerations are required, and this type of construction is most widely used by testing laboratories (see Fig. 6–4).

As implied above, one of the shortcomings of the cold-finger test is its duration, which is highly variable. This leads to conclusions about the nature of the deposits that are not justified based on the differences of shear and residence duration of field fluids versus those of the device. An additional complication arises when attempts are made to quantify the deposits produced. Reproducible recovery of the deposit from the probes, when run under the same conditions, is difficult and often leads to loss, as the time it takes to recover the deposit from the cold finger varies while the temperatures increases. Thus, the cold finger method is generally only semi-quantitative.

Cold-filter plugging test

Another method and apparatus employed for deposit testing is called the cold filter plugging test (see Fig. 6–5). This method and the various incarnations of equipment employed are also highly individual, and several different configurations exist. The cold-filter plugging test is concerned primarily with estimating the onset of crystallization, although it can be modified to include gross deposit tendency testing.

The concept employed by this method involves the idea that as waxes form, they can become impediments to the free flow of oils through restricted openings. Several devices are used to provide these restricted openings, and include: fine mesh wire screens, glass frits, formation core samples, special metal filters, and a host of different devices. These devices are generally submerged in the sample of oil, which is either cooled statically or dynamically, and a vacuum is applied across the filter. The time and temperatures are

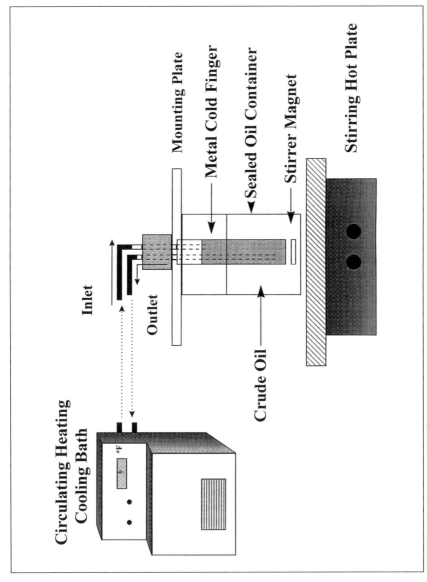

Fig. 6–4 Rudimentary cold-finger apparatus

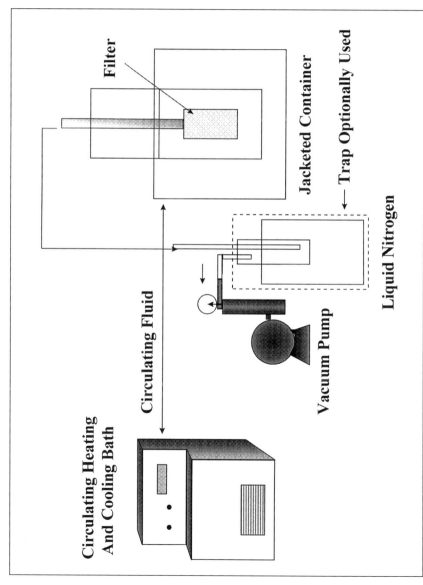

Fig. 6–5 Rudimentary design of cold-filter plug device

measured, and when the filter stops the flow of oil, this represents the plugging point, and the test is terminated. This method is often claimed to be sensitive enough to give accurate measurements of the cloud point temperatures of crude oils.

One of the obvious limitations of this test procedure is the complication introduced by sands, emulsions, and other discontinuities contained in oils. Another shortfall is that unless the filtrate is trapped in a cold trap, the quantitative utility of this test method is of no value. Test conditions must be accurately controlled so that temperature is constant or variable in a reproducible way, and vacuums must be carefully reproduced.

Dynamic test loops

Several dynamic tests and testing devices have been designed for use in the study of fluid pumping characteristic. These devices range in cost and complexity from thousands to hundreds of thousands of dollars, and it can take a large room to field portable units. Many employ computer-interface technology and cover a broad range in complexity of operation. Depending on what information is desired, these devices can use from a few to many sensors (or transducers).

These devices generally use pumps to circulate fresh (or recycled) crude samples through a conduit of specific dimensions, and measure such variables as temperature, fluid velocity, pressure, viscosity, pH, conductivity, water content, etc. The general concept of their operation is that by varying the conditions to which the sample is exposed (pumping rate, temperature, pressure, etc.) reliable data about the transfer of fluids can be obtained (see Fig. 6–6).

Rolled-ball deposit test

The rolled ball deposit test is among some of the routine testing methods used for chemical screening tests. This test consists of rolling a sample of solid paraffin isolated from the area under study (e.g., tubing, transfer line, or storage tanks) into a 3–4 gram ball. The ball is then placed in a container of water, solvent, or water-plus-surfactant, and after a quiescent period the extent of disruption is observed. The samples are then subjected to a period of agitation, and observed for the extent of disruption. The container and contents are generally duplicated, and one set of samples is heated, while the other is not, prior to the agitation. The intent of this test is to determine the effectiveness of various solvents and surfactants as deposit removal aids. This test is qualitative in nature and is subject to the individual prejudices of the one conducting the tests.

Dynamic rheology

Dynamic rheology has nearly become an indispensable tool in paraffin testing laboratories, because of its flexibility and sensitivity (see Fig. 6–7).

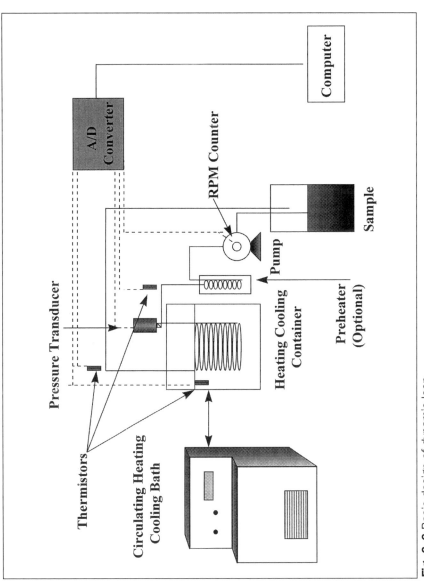

Fig. 6–6 Basic design of dynamic loop

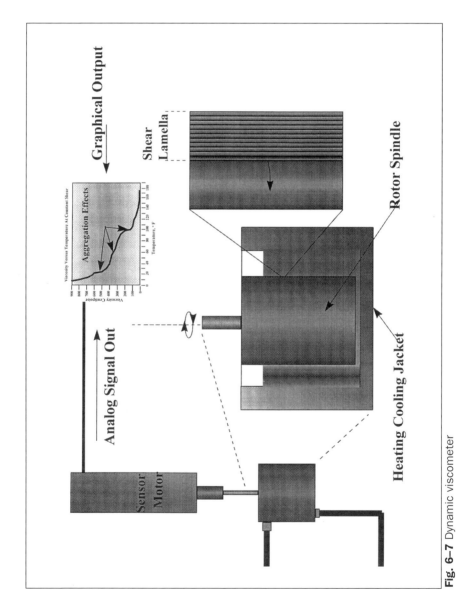

Fig. 6–7 Dynamic viscometer

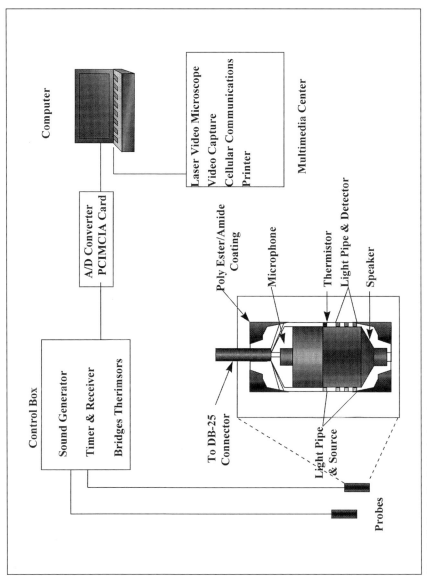

Fig. 6–8 Sonic Portable Laboratory

Waxy crude oils exhibit changing viscosity profiles as the temperature is changed. Thixotropy and yield point measurements are routinely performed as are other tests that seek to determine cloud points and temperature versus viscosity profiles. Several models of rheometers are available, and their costs range from a few thousand dollars to more than $100,000. The test methods performed range in complexity from simple temperature/viscosity profiles at standard pressure to high-pressure procedures performed in expensive closed systems.

Sonic testing

The Sonic Portable Laboratory is a device that consists of a collection of electromechanical sending and receiving devices that are interfaced to a computer through an analog to digital converter (A/D converter) (see Fig. 6–8). The concepts of operations performed by this unit are based on many of those employed by the devices mentioned above, with two notable exceptions:

- sound is used as a means of imparting shear forces to the sample
- a kinetic model is used in conjunction with temperature measurements to generate physical profiles of the system

Further, the device is designed to be portable and operate in a wide range of remote environments.

Because the unit employs probes for measurement, several measurements can be conducted remotely, under prevailing conditions of temperature and pressure. The unit is also modular, and can be fitted as part of some complementary package including such other devices as the dynamic test loop described above.

Supplemental test procedures

A test method has been described that utilizes a thin viewing window consisting of two plates of quartz, through which is circulated the crude oil under study, while a beam of polarized light impinges on the window. The transmitted light is then passed through a variable defraction grating on the opposing side, and the wavelengths are analyzed. The advantages claimed by those promoting this method are that the crude oil can be sampled under pressure from a producing well, or stem test, and maintained at pressure while circulating a sample between the plates of the cell. The cell can be heated or cooled, and it is also claimed that this technique is very accurate for determining the onset of a cloud point (see Fig. 6–9).

Summary

The preceding discussion gives an overview of some of the problems faced by the production, transport, and refining of waxy crude oils. One of the

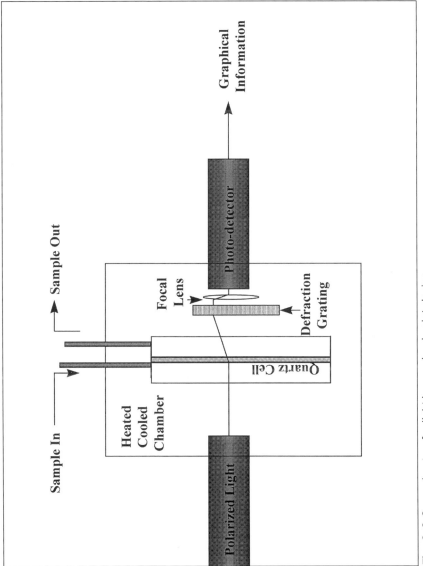

Fig. 6–9 General setup for light-beam cloud-point device

most important points to emerge from this discussion is the need for detailed planning. Planning by each facility from the producer to the refiner is necessary to avoid or remove the effects of wax accumulation. The discussion also points to the value waxes have to the end product mix, and why it is so important to provide adequate measures to afford its availability to the refinery.

Waxes and waxy crude oils present a challenge to those handling them, but they have value-added qualities that emulsion streams generally lack. The treatment of wax problems in crude oil is not always as simple as adding a crystal modifier. Sometimes it is necessary to resolve emulsions, disperse solids, or suspend asphaltenes before acceptable treatment is achieved. The physical conditions of the prevailing environment must also be considered; the best chemistry in the world will do no good if it cannot be pumped.

The preceding chapter also discussed some of the methods employed for the characterization of waxy crude oils. Each of these methods has its strong and weak points, but if they are carefully applied they can be of significant value to those involved in this work. Although several other methods are used in this work, most have not received the status or acceptance of those listed above. The extent of the problems faced by the producers, transporters, and refiners will determine the continued advancement of test methodologies.

Problems

6–1. What common forces of interaction between bipolar emulsifiers and waxes explain the increased difficulty of addressing either the wax problem or the emulsion problem?

6–2. What property of crystal formation phenomena explains the often-observed pour- point reduction by asphaltenes?

6–3. Will resolution of an emulsion in a wax containing crude always result in lower viscosity? Explain.

6–4. Would one expect the addition of an effective crystal modifier to a combination system of emulsion and wax to aid in the resolution of the emulsion?

6–5. If the highest fraction of wax in a crude oil melts at 60° C, would one expect to see a reduction in pour point by the introduction of a polymer with pendant groups that melt at 65° C? Explain.

References

Barrow, Gordon M. *Physical Chemistry*. 2nd ed. New York: McGraw-Hill Book Co., 1966.

Becker, H.L. "Notes on Chemical Test Procedures in the Oil Field." unpublished, 1983.

Handbook of Chemistry and Physics. 56th ed. Cleveland: CRC Press, 1975–1976.

7
Chemical Surfaces

Wax Surfaces

Chemical surfaces are very complex, and their study represents a vast area of interest and research effort. We can artificially divide this subject into two categories: *continuous* and *discontinuous.* The continuous chemical surface can be defined as a physical description of two differing materials that present the same physical appearance (e.g., liquid, solid, or gas). The discontinuous surface is then defined as a physical description of two differing materials that present a different physical appearance (e.g., emulsion or suspension).

This artificial distinction and the physical state represented are highly dependent upon factors such as temperature, pressure, and the degree of association. An example of the shortcoming of this artificial distinction is an emulsion of two substances that have the same refractive indices. Since they refract light the same, they present the appearance of a continuous and homogenous fluid. Our observation of this system is then subject to our perception of its visual appearance, and might lead us to conclude that the system is homogenous. Thus, it should be understood that discontinuities or continuities are more appropriately a reference to visually observable differences than composition. This implies that chemical surfaces should be looked at on the molecular level, and that the gross division of surfaces should be considered the result of the sum of the molecular surface divisions.

Wax aggregates exhibit some of the organizational traits of the surfactants that stabilize emulsions, since they tend to collect together and combine in groups of similar alkyl character (chain length). X-ray analysis of solid nor-

mal alkane crystals shows that the chains are extended, and that they form a zigzag arrangement comprised of numerous molecules aligned in an highly ordered array. This process of orientation in a zigzag pattern is continued until a significant depletion of the molecules forming the crystal has occurred, and the morphology displayed by these aggregates is in forms of zigzag platelets. This build up is continued, and aggregation by inclusion of smaller (lower melting point species) normal alkanes (paraffins) can occur to form bridged, higher-order aggregates (networks) (see Fig. 7–1).

Surface Hierarchy

In the examples of emulsions given earlier, the emulsion is described in terms of its gross presentation within the system. This description is adequate when we refer to its bulk behavior and appearance, but it should be recognized that an emulsion or wax crystal is a collection of several smaller structural entities. Thus, a water in oil emulsion consists of hydrogen-bonded aggregates of water molecules that contain solvated cations that also charge interact with carboxylate anions of the fatty acid.

These differing levels of interaction can then be considered as structural subunits of a higher ordered structure. Thus, anion-cation interactions can be considered as representing a secondary structural component. Surrounding each emulsion is a highly organized primary layer, with the polar heads oriented toward the center of the water ion mix representing a tertiary structural unit. A second layer of bipolar molecules is oriented about the primary layer, with the polar heads oriented outward to minimize the surface of the emulsion; this represents a *quaternary* structural unit. Finally a third layer of bipolar molecules forms a structure about the second with its nonpolar tails projected outward into the oil phase comprising a penternary (fifth-level) structure. This hierarchical structural scheme is repeated throughout nature, and is an indispensable part of the templates for biological viability (see Fig. 7–2).

The importance of analogous structures in nature cannot be overemphasized, nor can the power of analytical deductions made from them. A basic theme seems to emerge from the organization of complex structures in nature, and that theme is economy. Once nature develops a complex structure from less-complex starting materials, subtle changes accomplish dramatic gross structural differences. Because of this natural tendency and the relative ease of observing macro versus microstructures, most observations begin at the end and search for the beginning. Thus, our discussion of waxes follows this pattern.

Wax Defined

A practical definition of a wax might be that it is anything with a waxy feel and a melting point above body temperature and below the boiling point of water. Thus the term paraffin wax is used for a mixture of solid hydrocar-

Fig. 7-1 Paraffin crystal arrangement

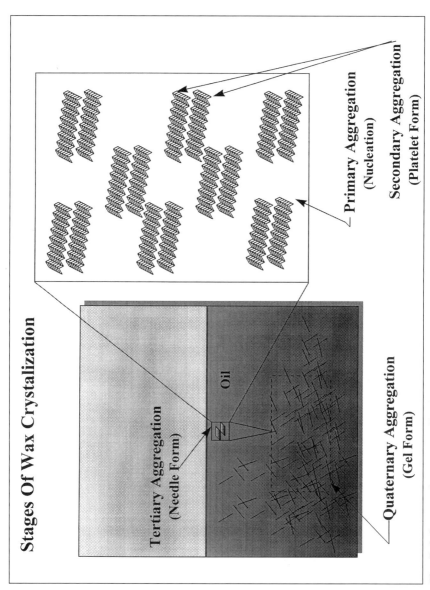

Fig. 7-2 Stages of wax crystallization

bons, beeswax for a mixture of esters, and carbowax for a synthetic polyether. Chemically, however, waxes have been defined as esters of long-chain (C_{16} and above), monohydric (one hydroxyl group), or alcohols with long-chain (C_{16} and above) fatty acids. Actually, the natural waxes are mixtures of esters and frequently contain hydrocarbons as well.

The definition of wax is extremely broad; fortunately, the waxes present in crude oil fit in a subcategory of this definition. Because of the anaerobic (lack of oxygen) conditions under which biodegradation of organic matter takes place to form crude oil, the majority of the waxes present are in the form of nonoxidized alkanes. These alkanes (C_nH_{2n+2}) may also exist as substituents of other complex organic components including aromatic, poly-aromatic, hetero-cyclic, and polymeric poly-sulfide parents.

An interesting feature of these complex materials is that the physical characteristics they manifest (e.g., viscosity or melting point) are largely determined by the length of the alkane substituents present. An example of this behavior is well represented by the synthetic polymers of octadecyl acrylate versus methyl acrylate.

While both polymers exhibit increased viscosity as the extent of polymerization increases (e.g., the length of the polymer backbone increases), only the octadecyl acrylate begins to cloud and solidify below 28° C. This clouding and solidification is due to the interaction of the $-C_{18}H_{37}$ pendent group, which has a melting point of 28.8° C. Therefore, when we refer to waxes present in crude oil, we are generally speaking about those complex components presenting physical characteristics that are consistent with those of the representative alkanes.

Wax and Viscosity

As alluded to above, viscosity effects can be thought of as arising from several different contributing mechanisms, and two important factors included in these mechanisms are polymeric and pendent group chain lengths (see Fig. 7–3).

The polymer backbone, which is covalently bonded (formally bonded, via inter-electronic energy level sharing), contributes to the bulk system viscosity and gives the system a high resistance to heat or shear thinning. While pendent group interactions arising from London forces of induction over van der Waals radii also contribute to the bulk system viscosity, these interactions are weaker and more susceptible to heat and shear thinning. The covalent backbone's contribution to the bulk systems viscosity is, therefore, considered irreducible under ordinary conditions of heating and shear, and is given the viscosity nomenclature of true Newtonian.

The pendent interaction viscosity effect is given the name thixotropic, which indicates that this viscosity contribution is susceptible to shear thinning. Although the London/van der Waals' interactions are weak, their greater number and high crystalinity combine to produce high viscosities at reduced temperatures (conditions favoring crystallization). As the concen-

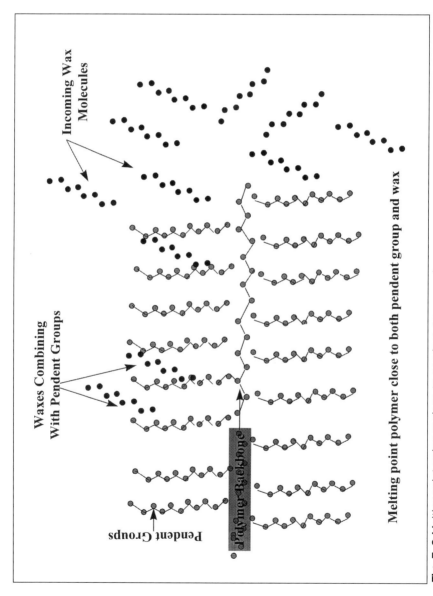

Fig. 7-3 Melting-point polymer close to pendent groups and wax

tration of these interacting species increase they can overwhelm the viscosity effects of the polymer backbone and become the dominant force determining the gross system's appearance.

The Purity Gradient

In the discussion of emulsions, we talked about the purity of the interface and how it affected the structural integrity of the emulsion. In that discussion we only touched on the reasons for the purity of the interface; now we will look a little more closely at these reasons.

Chemical structures exhibit patterns of solubility, melting point, and boiling point, which can be accounted for by their composition and external environmental conditions. Water boils at 100° C and freezes at 0° C when the external pressure is 760 mm of mercury (1 atm). However, if the pressure is less than 1 atm, water boils at a lower temperature, and conversely, if the pressure is higher it will boil at a higher temperature. At a point on a pressure versus temperature plot of water, all three forms of water coexist: gas, liquid, and solid. This point is called the triple point (see Fig. 7–4). This triple point is explained by a combination of several factors, among them hydrogen bonding, molecular velocities, and repulsive forces.

The degree and extent of hydrogen bonding determines whether water exists as a gas, liquid, or solid, and the extent to which hydrogen bonding occurs is determined by the allowed number of appropriately oriented water molecules (Fig. 7–5). The appropriate orientation for hydrogen bonding is determined by the geometry of the two molecules undergoing combination. The extent of combination, however, is determined by number of effective collisions. Velocity and positional duration of the interacting molecules must then be appropriate for the interaction to occur.

This situation suggests that the phenomenon of hydrogen bonding and the macro physical appearance of the system is quantized (only certain energy levels are available for interaction). The physical manifestations of water, as a consequence of the phenomena described above, can be extended to many of the behaviors of other chemical substances.

The physical behavior of bipolar materials, such as fatty acids, can be rationally explained in a similar fashion as that given above for water. Because fatty acids have a polar component carboxyl group (-COOH) and a nonpolar component R (where, R= C_nH_{xn+y}), two distinct physical interactions can be attributed to this structural arrangement. First, the possibility for hydrogen bonding exists where the partial charge on the oxygen of a nearby carboxyl interacts with the hydrogen of another nearby carboxyl group. Second, the R groups (alkyl groups) may also interact through weaker forces such as London forces of charge induction, and arrange themselves as in the illustration of the emulsion structures shown earlier.

We now see that an additional complexity has arisen from the introduction of the R group. The fatty acids exhibit solubility, boiling point, and melting points just as water does. However, the fatty acids have an additional

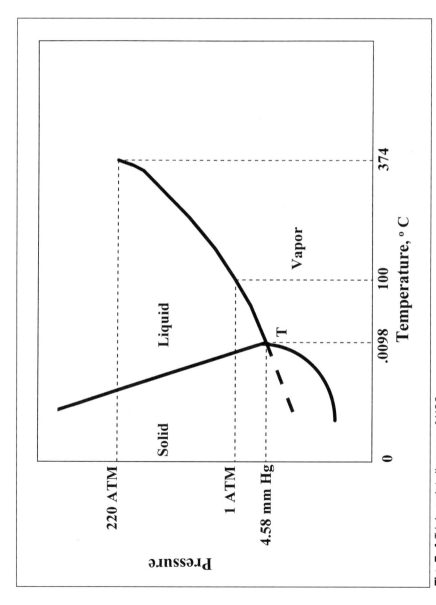

Fig. 7–4 Triple point diagram of H2O

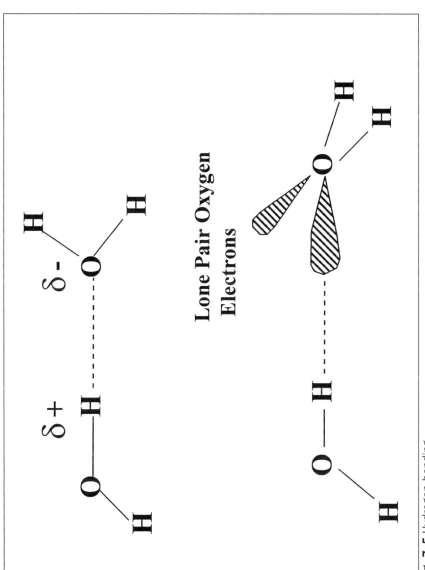

Fig. 7–5 Hydrogen bonding

facility for physical interaction that results from the presence of the nonpolar R group.

Combining water and fatty acids now begins to take on additional meaning. If a fatty acid is added to water with very little turbulence, the less dense fatty acid orients its polar head (carboxyl groups) to the water, and its fatty tail toward the air above (see Fig. 7–6). As turbulence increases, either through heating or mechanical mixing, emulsions are formed wherein the fatty acid orients itself to minimize its exposure to the water surface (see Fig. 7–7). This behavior helps to explain the properties of a dual component emulsion, but emulsions are generally more complex than this, and often contain several emulsifying agents.

Adding a second fatty acid to the system will now comprise a three-phase system. The first fatty acid will now be displaced by the second, and a concentration gradient will form between the two. The gradient will be determined at the surface by the size and configuration of the R groups and their affinities for one another (like dissolves like). Now when agitation is applied, the emulsions that form will consist of interfaces that represent a purity gradient of the two fatty acids present in the system.

Thus a partition has developed arising from the favored tendency of like group association. This partitioning results in a distribution of emulsions sizes that are divided into roughly dual radii on average, and the bulk physical properties of viscosity will reflect this character.

Wax and Changing Surfaces

The composition of crude oils includes several bipolar species capable of forming emulsions, and the variation of the polar and nonpolar portions of these species determines the variety of the emulsion sizes present. The differences in chemical functionality (e.g., hetero-cyclic amines, sulfides, oxides, carboxyl, poly- and mono-aromatic, alkane, and alkene groups, etc.) and the attendant separations between emulsion stabilities tend, therefore, to be broad.

The situation with waxes and their structural make up presents a more diffuse picture. There are several reasons for this diffuse physical behavior, but two of the most important are the subtle chemical separation between the interacting species and the differences of the interface presented to the external phase. While emulsions present a sharp phase distinction over a broad range of temperature, waxes present a range of physically distinct phases from solid to liquid.

Surface Tension and Wax

The physical behavior of wax-forming species can be described in a way that is consistent with the physical behavior of water (e.g., the triple point). Although the melting and boiling points of wax-forming species are generally much higher than that of water, the type of forces responsible for their interactions are much weaker. The boiling point of a substance can be viewed

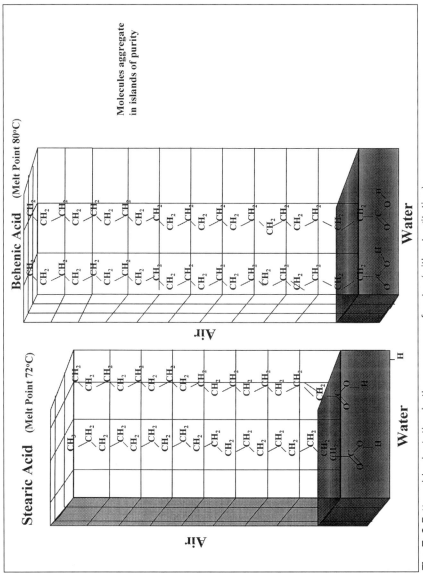

Fig. 7–6 Fatty acid orientation in the presence of water (without agitation)

135

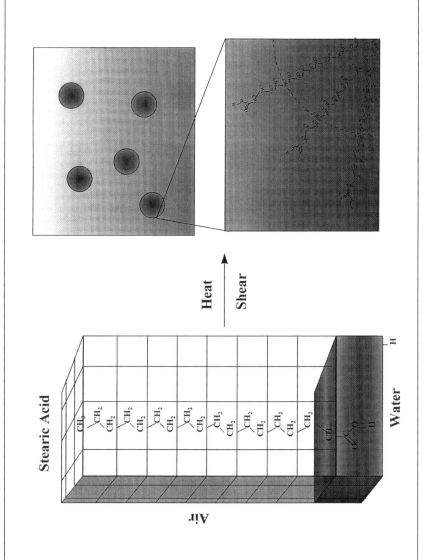

Fig. 7–7 Orientation of a fatty acid in a water emulsion

as the attainment of a molecular velocity sufficient to overcome the surface tension barrier of a system. As such, molecular speeds play a major role in the appearance of the bulk system.

Maxwell's equation for molecular speed distribution shows that this distribution is solely a function of temperature. The following formula shows James Clerk Maxwell's derivation describing the most probable speeds of a large number of gas molecules:

$$N(v) = 4\pi N(m/2\pi kT)^{3/2} v^2 e^{mv/2kT}$$

where

$N(v)dv$	is the number of molecules in the gas sample having speeds between
v	and $v+dv$.
T	is the absolute temperature
k	is Boltzmann's constant
m	is the mass of the molecule

Thus the boiling point of a simple hydrocarbon like heptane (C_7H_{16}), being around 98° C, indicates that at this temperature its surface tension can be related to that of boiling water.

This relationship then suggests that, at their boiling points, both heptane and water possess a molecular momentum (mass x velocity) that is large enough to overcome the surface barrier forces acting to prevent their escape. The mass of heptane is 100.21 g/mole, while the mass of water is 18 g/mole. And, because speed distributions are a function of temperature alone, the cohesive force per unit mass at the surface of water must be a close to ratio 100:18, or about 5.5 times greater than heptane. The reported values for water and hexane are 80 and 18 dynes/cm^2 (or 4.44:1) at 20° C, respectively (*Handbook of Chemistry and Physics*).

By employing the first principles, as above, a rough estimate of the surface tension difference or intermolecular forces of association was obtained. Having gone through this exercise, an enhanced understanding of the intermolecular forces acting to allow molecular association can be achieved. The phenomenon of boiling represents a physical change from liquid to gas, while the phenomenon of melting represents a physical change from solid to liquid. Each of these phenomena are truly representations of the degree of association or conversely the extent of division of the components comprising a system.

Thus, the transition of water or hydrocarbon from liquid to gas is a measure of the division of the molecules. The strength of the interaction between these molecules is then a measure of the forces that act in opposition to this division. Because velocity is intrinsically related to each of these phenomena, and the distribution of molecular velocities is solely a function of temperature, it must be concluded that not only is velocity directly proportional to temperature, but surface tension is inversely proportional molecular velocity.

The ramifications of this inverse proportionality are not as obvious as they may appear, since we are talking about a population of associated and disassociated molecules exhibiting a range of velocities. Consequently, the bulk physical appearance of the system is as much a function of the number of molecules populating a particular velocity as the momentum possessed by the molecule (or molecular aggregates) at a particular velocity.

Considering this relationship helps to explain how secondary, tertiary, quaternary, and higher structural aggregates arise. It is clear that chemical systems tend to organize in patterns, and that these patterns are determined by their chemical similarities. These similarities are represented by chemical functionality and the forces derived from their interactions. Further, state functions such as internal energy act to produce conditions that determine the population of species within a given range of momenta.

Wax Crystals

Having laid some groundwork for the discussion of the phenomenon of structural aggregates, it is now appropriate to begin an examination of one of the most common, yet least understood, members of this group. Organic crystals are ubiquitous in nature, and range from highly functionalized (e.g., carboxylic, phenolic, aromatic, hetero, and homo-atomic, polymeric, mono- and poly-cyclic groups etc.) to monofunctional substances (e.g., normal paraffin) (see Fig. 7–8).

The class of organic crystals represents a broad range of geometries, including needles, plates, cubes, rods, prisms, pentagons, octagons, hexagons, rhomboids, and pyramids. Each of these forms result from crystallization from a solution. The geometry of the crystals formed is determined by the solute/solvent interaction, which we now understand to be a function of the composition of the solute, solvent, and the physical conditions of the system (e.g., temperature, pressure, and mechanical mixing).

The physical conditions, however, are justifiably lumped into a thermodynamic function commonly known as entropy (a measure of the randomness of the system). This measure of randomness or entropy is quantifiable in terms of the entire system, and therefore represents a gross measurement of the summation of the various individual interactions. It should be noted that because we are dealing with dynamic variables such as velocity, momentum, and energy, time is intrinsic to this discussion.

This intrinsic quality of time is generally viewed in terms of the duration of a physical phenomenon, such as distance at a velocity. However, this concept demands duration be a function of an arbitrary unit of measurement marked off by an artificially defined unit or second. As we will see, physical systems tend to have a different interpretation of time.

Crystal Order

As indicated in the preceding discussions, the structural aggregates comprising crystal forms owe their morphology to their individual chemical

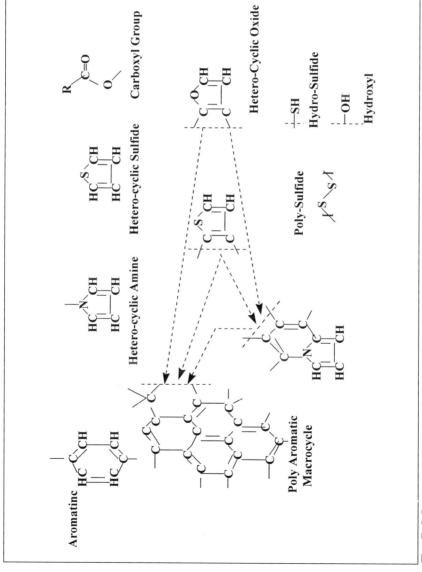

Fig. 7–8 Some complex chemicals found in crude oil

functionality and the entropy of the system. The matter of functionality is quite easily understood by differences in polarity arising from hetero-atomic nature of these functional groups. It is quite a bit more complicated to understand the reasons for the variability in geometry exhibited by chemically identical functional materials that form from identical solvent systems.

This is where entropy comes into play, and in order to understand entropy's effect on the morphology of the crystals, we must introduce the concept of intrinsic time. Although this concept may seem foreign to the reader, it is commonplace in nature and is at the seat of all observation; time is a dimension. Time must then be considered as valid a part of the domain occupied by a growing crystal as the x, y, and z components that occupy space. This stipulation makes it possible to discuss the root causes of differing crystal morphology.

We have discussed the sole dependence of the distribution of molecular velocities on temperature of the system, and we have also discussed the concept of the population of species that occupies specific velocity ranges. The role of time was implicit in these discussions, but remains ambiguous. Let's examine this role with respect to population and temperature.

If we look at a plot of the number of molecules/given-velocity versus velocity at two different temperatures, we see that the curves are both bell-shaped, but that the curve for the lower temperature is more narrow than that of the higher temperature (see Fig. 7–9). This fact has a very important consequence when we try to reconcile the reasons for morphological differences between two chemically identical, but physically different, systems.

In order for two molecules to combine to form an aggregate or distinctly different substance, an energy barrier must be overcome. The process of surmounting this energy barrier requires at least two criteria be met: proximity and proximity duration. Both of these criteria are functions of velocity and therefore temperature, thus only the populations of molecules occupying a velocity range that is neither too great nor too small will combine.

A more succinct way of stating this would be to say that only appropriately quantized energy states can combine. Thus, the momentum necessary to penetrate the energy barrier of repulsion must not only be sufficient to place the interactive species in close proximity, it must also maintain this proximity for a sufficient period. Or mathematically this can be written as (energy x duration)/momentum = minimum interactive distance.

Crystal Order and Surface Tension

The previous discussion of surface tension described the liquid/air interface, and its analogy to physically heterogeneous surfaces. In this section an effort will be made to describe how the surface of a pure solid interacts with its pure liquid form to produce crystals of differing morphology. From the discussion of functionality, there would appear to be no reason for heterogeneous crystal formation since we are discussing pure systems. But given the discussion on velocity and more particularly on temperature, we can begin to understand the phenomena leading to differing crystal forms.

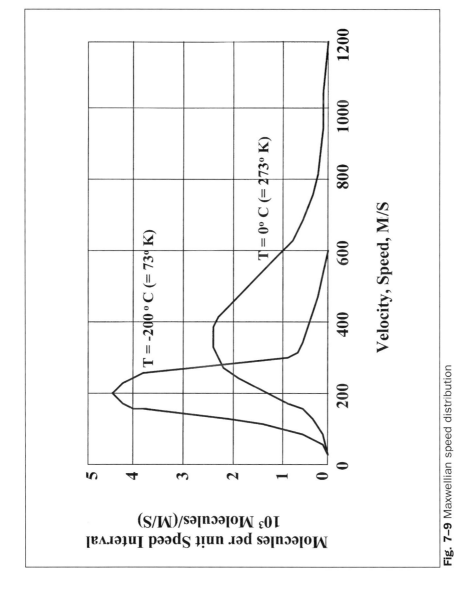

Fig. 7–9 Maxwellian speed distribution

141

When two physically different forms of chemically identical materials coexist in suspension form, as does ice in water, the interface is in a constant state of flux. Liquid water exchanges places with solid water (ice) at the interface, and vice versa. This continuous interchange renews (reforms the shape) the entire solid liquid interface over time, assuring a clean and reactive surface while changing its morphology. The same is true for organic liquid/crystal surfaces, and the operative mechanism for this interchange is facilitated by escape velocity versus surface tension between the two surfaces.

The liquid water possesses sufficient energy to escape its surface (the energy barrier envelope surrounding the crystal). Meanwhile the ice is transformed to liquid by the excess momentum possessed by entering liquid water and thus achieves sufficient energy to penetrate the surface in the opposite direction.

The strength of the forces that hold the crystal in the solid state are therefore dependent upon the combination of inter-atomic association, or surface tension. Each form of crystal then has associated with it a vibrational component between combined molecules that arises from external momenta, and these vibrational components tend to be uniform within the crystal, but nonuniform at or near the surface. Thus, both the energy- barrier resisting combination and the vibrational energy of the crystal represent time-dependent phenomena.

One of the clearest indications that these state changes are quantized is the phenomenon of *sublimation*—the process of changing from solid to gas without passing through an intervening liquid state. Solid carbon dioxide (dry ice) sublimes or transforms directly from solid to gas without passing through the liquid state under conditions of standard pressure and temperature (1 atm 25° C). Thus, time is an inseparable part of quantized transitions, since temperature is solely a function of velocity, and velocity can only be described in terms of passing time or time intervals.

Summary

Phases are artificially classified as continuous and discontinuous, but these classifications are more a matter of convenience than accuracy. This artificial distinction results from the desire to see immiscible substances as distinctly demarcated by a sharp boundary. Higher-ordered structural systems result from partitioning effects (e.g., solubility differences, charge interactions, and inductive forces). Waxes form higher-order structures through solubility and inductive forces, and the degree of structural complexity is a measure of their concentration and individual molecular weights. It is possible for multiple physical states (gas, liquid, and solid) to coexist under particular conditions of pressure and temperature.

Paraffins exhibit melting points that are consistent with the longest uninterrupted linear carbon-carbon segment. Polymers containing pendent paraffin substituents exhibit melting points that are near the melting point of the pendent substituent. Viscosity effects are contributed by the concentration, molecular weight, and functional group interactions between components within a system. These effects can be due to strong forces (e.g., covalent polymer bonds, ionic interaction, hydrogen bonding, etc.) or they can be the result of combined weak force effects (van der Waals inductive forces). The treatment given was intended to show the relationship between statistical systems and their physical manifestations.

Velocity, populations of interacting materials, and the duration of proximity within interactive domains were also discussed.

Problems

7–1. Explain why velocity distributions in aggregate systems, such as wax in oil, behave in an analogous fashion to gases.

7–2. Describe how surface tension varies with temperature.

7–3. What relationship to molecular or aggregate velocity does hierarchical arrangement manifest?

7–4. Discuss velocity distributions of aggregates in relation to London forces and van der Waals radii.

7–5. Describe how duration at van der Waals radii affects the aggregation of molecules.

7–6. Discuss the shortcomings of the artificial distinctions of discontinuous and continuous surfaces in the description of aggregate mixtures.

7–7. Hierarchical structures present a basic theme in nature and are common in biological systems. What is that theme? Present some ideas why it is observed in so many complex systems.

7–8. Give a practical definition of wax.

7–9. Given a polymer of several thousand molecular weight consisting of a polyethylene backbone, and octadecyl pendent groups, which melting point temperature would you expect to observe, that of the polyethylene backbone or the pendent group?

7–10. Is it necessary for a molecule or aggregate of molecules to possess a dipole moment in order to induce charge in a neighboring molecule or aggregate?

7–11. Explain why, if ionic interactions are stronger than London forces, bulk fluid behavior is often determined by the latter rather than the former.

References

Barrow, Gordon M.: *Physical Chemistry.* 2d. ed. New York: McGraw-Hill Book Co., 1966.
Handbook of Chemistry and Physics. 56th ed. Cleveland: CRC Press, 1975–1976.
Noller, Carl R. *Textbook of Organic Chemistry.* 3d. ed. Philadelphia: W.B. Saunders Co., 1966.
Resnick, R., Halliday, D. *Physics, Part I.* 1st ed. New York: John Wiley & Sons, Inc., 1966.

8
Wax Crystal Order
and Temperature

The Odd Relationship
of Time and Temperature

Certain implications regarding temperature and time emerge from the view so far expressed. These implications suggest a dependency of time on temperature and vice versa, and that the population of materials within a velocity range is integral to the temperature. Thermodynamists have noticed this curious relationship and have hypothesized that temperature is the complex conjugate of time, which can be expressed in the form of (a+ib) where $(-1)^{1/2} = i$. This view can be considered to mean that *temperature is imaginary time*. Thus a two-dimensional plot of this imaginary surface will have an origin corresponding to very hot and very early and diverge at right angles toward very cold and the present (see Fig. 8–1).

Crystals, perhaps more clearly than other forms of matter, add an intuitive insight to this complex relationship, since they represent a very pure form of creation of order from the midst of chaos. The anti-entropy behavior exhibited by growing crystals can be thought of as an isolated battle against the natural tendency toward a maximum of disorder. Or it can be thought of as a little arrow of time opposing the overwhelming cumulative arrow that points to a maximum of chaos.

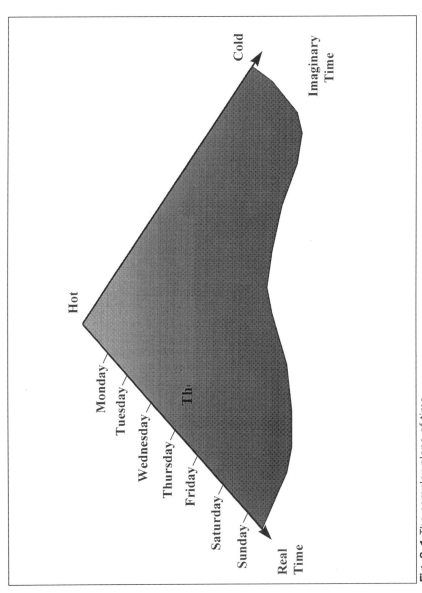

Fig. 8–1 The complex plane of time

Wax Crystal Habits

Thus far the discussion has been confined the general behavior of complex aggregates with an emphasis on crystals. An effort has also been made to explain the general nature of the interactions that account for the existence of these complex structures. The method of development and order of these discussions were intended to develop a degree of familiarity, allowing us a means to increase our focus on some specific types of crystal behavior. Particular types of aggregates such as paraffin crystals paradoxically represent a high degree of complexity as a result of their structural simplicity.

The paraffins possess a single chemical functionality that has no inherent charge or polarity since they consist of only carbon and hydrogen. Therefore, their aggregate association is accomplished through high proximity and charge induction (van der Waals radii) that have overcome dispersive forces of attraction and repulsion (London forces).

London Forces and van der Waals Radii

London or dispersive force arise from the relative positions and motions of two molecules. When these molecules approach one another, their individual envelopes of intra-atomic molecular orbital configuration are distorted. Each molecule resists this distortion resulting in an increasing force opposing continued distortion, until a point of proximity is reached where London inductive forces can come into effect. If the velocities of these molecules are high enough to allow them to approach another molecule at a distance just equal to the van der Waals radii, they can combine (see Fig. 8–2).

The continued distortion of the intra-atomic orbital configuration results in a bimolecular arrangement or interatomic orbital configuration that has sufficient proximity to blend the electronic orbitals of one molecule with the other. This blending of molecular orbitals produces an induced dipole, or statically charged couple that now resists separation. The dipole forces are called London dispersion forces and represent the weak forces responsible for like molecular aggregation.

As in previous sections, the forces acting to accomplish a higher order structure are opposing the forces acting to prevent this structural aggregation. The difference is the strength of the aggregation force, and these forces are weak compared to hydrogen bonding or charge interactive forces (see Fig. 8–3).

However, what these weaker forces lack in strength is often compensated for by the number of molecules present undergoing these interactions. Refer to appendix C for a more detailed discussion of these phenomena.

Molecular Crystals

The relationship of time, temperature, and the quantum nature of physical interactions had to be brought into focus before a discussion of the

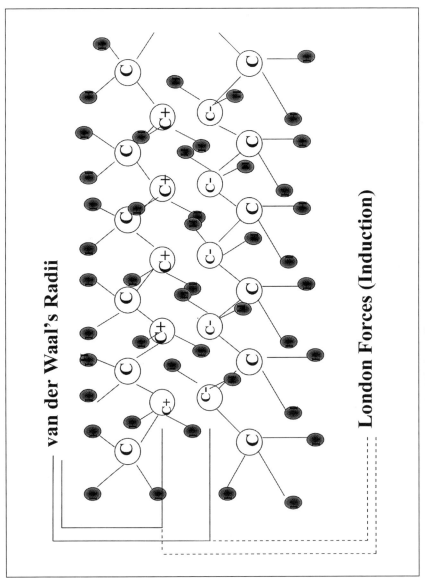

Fig. 8–2 London forces and van der Waals radii

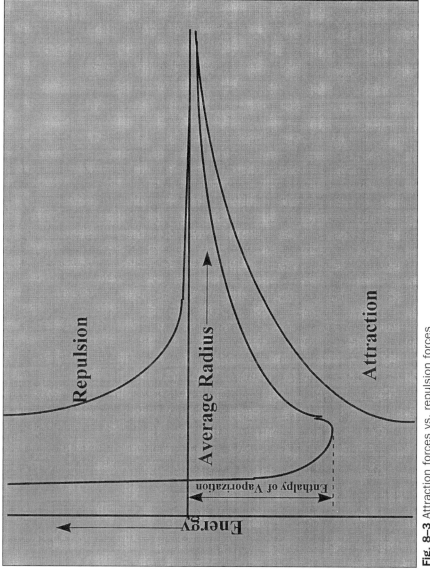

Fig. 8–3 Attraction forces vs. repulsion forces

following empirically derived model of wax crystal behavior could be properly discussed. When crystals form from a solvent, heat is evolved from the forming crystal and transferred to the solvent. This transfer of heat results from the radiation of excess inertial energy from the crystal side of the energy barrier as proximity and interatomic combination is achieved.

Taken individually the heat quantity transferred is very small per interaction, but in systems with large numbers of forming crystals, the cumulative heat quantity is measurable. This radiant heat transfer then adds to the intrinsic solvent heat, thus slowing the measured cooling rate of the mixture. Therefore, if a pure solvent and a mixture of that solvent and solute are both cooled at a specific rate by an external cooling system, the pure solvent will cool at a faster rate than the mixture. The variance in the cooling rate between the pure solvent and the mixture will, therefore, be a function of the solute composition and concentration.

When studied in terms of the effects of solute/solvent mixtures versus freezing point temperature, a well-known phenomenon called molal freezing point depression is observed. This phenomenon and its attendant changes to the system can be used to calculate the molecular weight of the solute in dilute mixtures (see Fig. 8–4).

Consider a two-component system, at some fixed pressure, where the temperature range treated is such as to include formation of one or more solid phases. A simple behavior is shown by those systems for which the only solid phases that occur are the pure crystalline forms of the two components. Such phase behavior is exhibited by a mixture of benzene-naphthalene. It is informative to consider what happens when solutions of various concentrations are cooled. The data that are obtained give the temperature of the system as a function of time. These data are plotted as cooling curves, and it is such cooling curves, in fact, that are used to obtain the data shown in the phase diagram (see Fig. 8–5).

Ternary and Higher Eutectics

Binary eutectics can be graphically depicted as bimolar component concentrations versus temperature at constant pressure. However, when ternary eutectics are plotted, pressure and temperature are held fixed to allow a construction of a two-dimensional diagram. Thus, a triangular plot is used to represent the behavior of a three-component system (see Fig. 8–6).

The corners of the triangle represent the pure component, and contour lines are constructed to represent the various blends. According to Barrow, "A three-dimensional representation...shows, in a descriptive manner, the phase behavior as a function of composition and temperature at the fixed pressure of 1 atmosphere." Extreme graphical difficulties arise when more than three components are considered, and the adequacy of quantitative three-dimensional diagramming becomes cumbersome. Remember, however, that the data used in the construction of the phase diagrams is obtained from

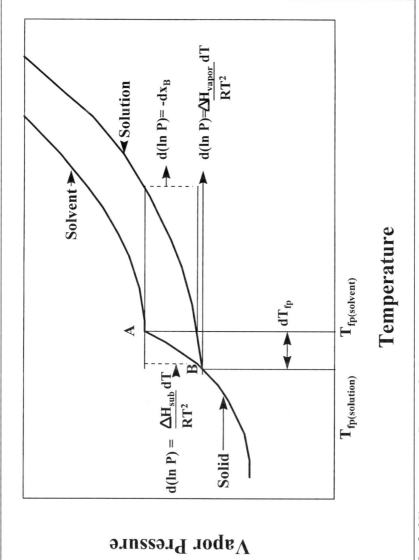

Fig. 8–4 Vapor pressure diagram of solvent and solute near freezing point (at 1 atm)

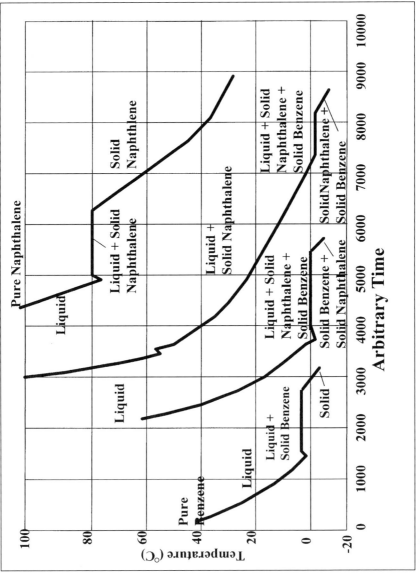

Fig. 8–5 Cooling rates of naphthalene/benzene mixtures

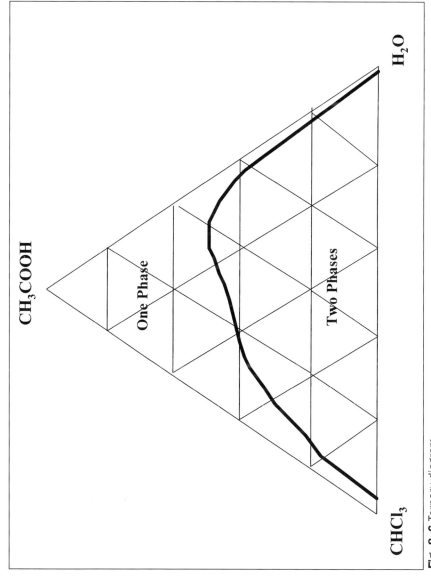

Fig. 8–6 Ternary diagram

the cooling rate curves. Considering this fact, might it be reasonable to use these cooling rate plots for the quantification of multiple component systems? This is the route taken in the development of a kinetic model of pour-point behavior(see Fig. 8–7).

Examination of the cooling rate curves depicted in the example of the binary mixtures of benzene and naphthalene reveals temperature plateaus. These plateaus are reflections of phase transitions that take place in the mixture. Thus, the period of time these plateaus persist (e.g., temperature remains constant while time passes), should provide information about the mixtures composition. If a mixture is subjected to carefully controlled external methods of heating and heat removal, then the change in the inflection of mixture temperature versus time should provide information about the mixtures composition.

It should also be recognized that the curve of temperature versus time is a running summation of the individual crystal contributions to heat evolution, reflected by a concomitant increase in temperature of the overall mixture. Thus, the area under the cooling curve represents an integral form of a differentiable function, or the equation for the curve. If distinct equations for mixture cooling rate curves can be derived for binary systems, then it should also be possible to derive equations for ternary and higher mixtures.

Introduction to a Kinetic Model

Observations of the cooling rate behavior of complex-wax-containing crude oils treated with crystal modifiers versus untreated samples showed the following: Successfully treated (crystal-modified) samples cooled at a more rapid rate than untreated samples. Numerous inflections (temperature plateaus) were observed. The duration of the temperature plateaus and the temperatures at which they occurred varied for different crude oil samples, and for treated versus untreated samples. The duration of the plateaus, in all samples, was inversely proportional to the temperature difference between the mixture and externally applied temperature. The results of these observations, with the exception of the accelerated cooling rate of the treated sample, are not surprising given the above discussion.

What happened to the mixture when the crystal modifier was added? And what changes were affected by the modifier's addition that caused the mixture to cool more rapidly than its untreated counterpart? In order to answer these questions, it is necessary to draw on the information presented in the sections preceding this discussion. If we focus on a mixture consisting of three components (solvent, solute, and crystal modifier), realizing that the solvent, solute, and crystal modifiers also consist of several components, we get a feeling for the complexity of these questions. If, however, we set criteria for the molecular interactions as described above and limit our focus to the general phenomenon of interaction, some clarity can be achieved.

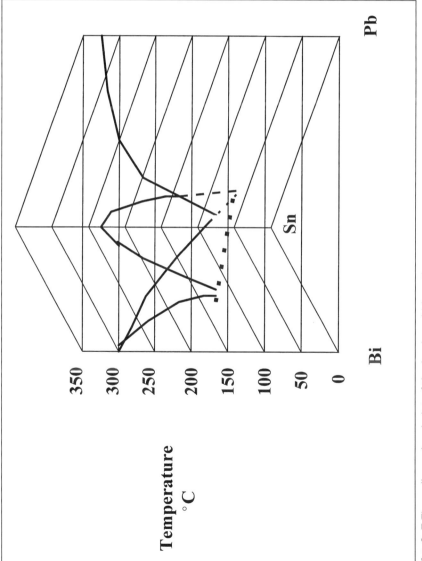

Fig. 8–7 Three-dimensional plot of tin, lead, and bismuth

Crystal Modifiers

Paraffin crystal modifiers are a broad class of chemically functionalized materials. Their structures range from poly-acrylate esters of fatty alcohols to copolymers of ethylene and vinyl acetate. These chemicals are so constructed that portions of either the backbone of the polymer, or their pendent groups, interact with the crystallizing waxes present in a crude oil mixture.

The nature of the interaction of these polymers with the wax-forming molecules in the oil determines their effectiveness as modifying agents. If we think of the formation of the wax from the point of view of group association effects, and the wax's interaction with the modifier similarly, molecular and aggregate velocities must be considered. Consequently, the mechanistic interpretation of crystal modification must account for velocity (momentum) effects that are consistent with observed behavior.

Mechanistically the momentum, and therefore the energy inherent to a molecule, is a function of its velocity and acceleration. But we have seen that the velocity distribution of a system of molecules is determined solely by the temperature of the system (Maxwells' distribution of molecular speeds). This means that only the average velocity of a system of molecules at a temperature is immediately obvious from the temperature. Thus, there is a substantial population of molecules with different velocities, accelerations, and energies at either end of the mean population distribution.

Combining this condition with the criteria of quantization suggests that populations of wax-wax and wax-polymer aggregates will partition within velocity ranges. Thus, by the application of first principles, a consistent picture of the nature of interaction of the polymer with the wax is obtained. Although this is a good start to understanding the mechanism of crystal modification, it is only a start. There are still several questions about the subsequent interactions that need to be addressed. Now that a strong case for incorporation of the crystal modifier at the site of wax nucleation has been developed, the subsequent behavior of the modified wax must be examined.

Just as polymer crystal modifiers are incorporated into aggregates of wax, so too are higher molecular weight waxes, since they occupy several velocity ranges. This apparent complication, by its existence, acts to elucidate the nature of the mechanism of interaction. The reason it does is that it points to a specific interaction temperature range (velocity population) wherein the polymer wax interaction is operative. Therefore, the successful depression of pour point (successful crystal modification) is not solely dependent on the necessity of interaction of the polymer with the highest wax species present. Rather, it is more important for the polymer to interact with that wax fraction representing the most highly populated velocity partition at a given temperature.

By so doing, the surfaces presented by the growing crystal are altered, and the energy barrier is changed. This change also affects changes in the required inertial forces to overcome the barrier, which after alteration disallow interchange and residence in proximity for sufficient time for interactive forces of attraction to be established. This mismatch results in a greater distribution of velocities external to the growing crystal, and as a consequence increases the radiant energy dispersal to the environment (the system cools faster).

A Return to a Kinetic Model

The prospect of using the cooling rate curves exhibited by complex systems, like crude oils, to derive information about the components of the system was examined. Thus, a series of pure normal paraffins (99.9%+ pure) ranging from $C_{16}H_{34}$ to $C_{36}H_{74}$ were blended in various proportions as single and multiple components in a narrow distillate cut of kerosene. These blends were then subjected to various conditions of cooling, and curves of their temperature versus cooling time were plotted. The aim of this study was to confirm the predicted behavior of these systems, and to derive an empirical mathematical fit to the cooling rate curves of the blends (see Fig. 8–8).

The correlation of known variables to the cooling rate profiles required very tight control of the external temperature, and a highly sensitive method of measuring the internal temperature. This was accomplished by proportional heating and cooling of the external cooling bath, and sensitive thermistor readings of both the sample and external bath temperatures. The following empirical formula was developed from these studies:

$$T_{mp}^{2}/C\#^{log\,C\#} = k\text{'}dt$$

where

T_{mp}	is the melting point temperature ($^\circ$ K) of the pure normal paraffin
$C\#$	is the carbon chain length
dt	is the elapsed cooling time in seconds
$k\text{'}$	is the constant of proportionality ($^\circ K^2$/sec)

(Note: the discussion of the relation ($iT = t$) and the units of the ratio (h/nk) in Appendix C might now be used to fully appreciate $k\text{'}$.)

Once established, this relationship combined with the realization that the curves represent a running sum of the individual contributions to the heat evolved afford a method of deriving information on the composition of the system. Thus, the integral of the curve represents a summation of the individual component contributions to the heat evolved over a cooling period. Therefore, subtraction of the preceding sum from the succeeding sum should yield a measure of the component present as illustrated in the following:

$$100\text{-}\{ k\text{'} [Ú\,n\,dt - Ú(n\text{-}1)\,dt]/[Ú\,mT\,dT\text{-}Ú(m\text{-}1)T\,dT]\} \propto Area\ \%$$

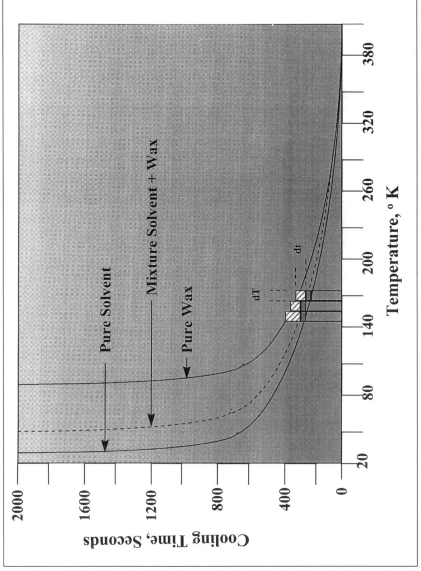

Fig. 8–8 Kinetic model cooling rate curves (dT/dt)

where

n and m represent corresponding time and temperature
points on the curve, which are incremented by the
next correspondence values of time and temperature

Setting the system such that it always represents 100% then yields differential values, which are normalized for the component composition (see Fig. 8–9).

Additionally expressions for k' and $C\#$ (carbon chain length) can be derived from the integral curve expression as follows:

$$k' = (log(T^2/dt))^{1/2}/((dt/T^2) + 2.303)$$

where

T is the measured sample temperature ($^\circ$ K)
dt is the measured cooling time (sec)

The carbon chain length is calculated as follows:

$$C\# = 10^{k'(dt/(T \times T)) + 2.303)}$$

where

$C\#$ is the empirically calculated carbon chain length

Thus, the calculation of these values from the experimental data is used to develop a composition profile of the complex system.

Another extremely useful by-product of the investigation of the kerosene/wax studies was the development of a database of synthetic starting raw materials used to produce polymeric crystal modifiers. Because of the algorithms developed in the kinetic model technique, and the predictable effects attributable to alkyl group substitutions, it was possible to build a database of mathematically permuted combinations of raw materials. This database can be searched using profiles obtained from crude oil samples, and a prospective treatment chemical predicted by comparisons of the crude profile with a mathematically permuted raw material profile. Very good approximations are possible using this method.

Gas Chromatographic Composition Profiles versus Kinetic Model

When multiple component mixtures are subjected to gas chromatographic separations, the separations are accomplished by measuring the retention time on a column swept by an inert gas (helium or argon) while raising the

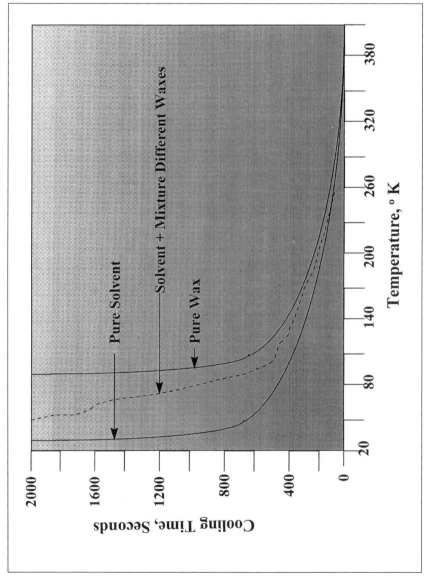

Fig. 8–9 Kinetic model cooling rate curves

temperature of the column to volatilize the components injected onto the column. Thus, the property being examined is boiling point, and as such intergroup associations being measured are most indicative of strong interactions, while the weaker interactions like the London van der Waals are overwhelmed.

Cooling curve measurements, however, are mild, and the manifestation of the resultant forces of these interactions are more likely to be detected. Thus, component profile comparisons between these different measurement methods should not be expected to produce very good correlation (see Fig. 8–10). Nevertheless in systems with low percentages of polar groups or other strongly interacting components, the correlation is quite remarkable.

As with most methods of analysis, each has its strengths and weaknesses. Gas chromatography (GC) is highly accurate and requires only micrograms of sample to produce dependable spectra. GC measures boiling point profiles, and consequently the stronger forces of interaction (polar interactions), while overshadowing weak interactions. However, this overshadowing of weak intermolecular forces then leads to a lumping together of molecules of highly different structure (polar, nonpolar, multifunctional, monofunctional, structural isomers, etc.) into boiling point categories.

The kinetic model method requires a substantially higher quantity of sample (usually 10 g or more) to give a dependable cooling rate profile. However, because of much milder conditions of measurement, it is capable of measuring the weaker forces of interaction. The kinetic model method also makes little distinction between strong interactive forces and weak forces, since it is a measure of bulk system behavior. However, the kinetic model method does distinguish between structural isomers. It makes sense to use both methods, therefore, to produce a more complete picture of the system.

Summary

The foregoing chapter discusses some of the physical phenomena associated with crystallizing materials. The major emphasis was placed on the physical manifestations of heat evolution, physical states, and combinations of these manifestations. The topic of eutectics was discussed, and in this connection the development of a "kinetic model" was presented. Additionally, the forces involved with crystal modifier interaction were introduced to prepare the reader for the chapter on testing methodologies.

Problems

8–1. Give an explanation of why plateaus of temperature are observed in curves of solute/solvent mixtures when temperature is plotted against time.

8–2. Construct two 3-D graphs of a three-component system, using temperature and time. What curve surface differences would you expect to observe in the time plot versus the temperature plot?

8–3. Describe the different effects in cooling rate behavior of a system composed of 30% octadecane in kerosene versus a mixture of 15% doecosane, 15% octadecane, and kerosene mix.

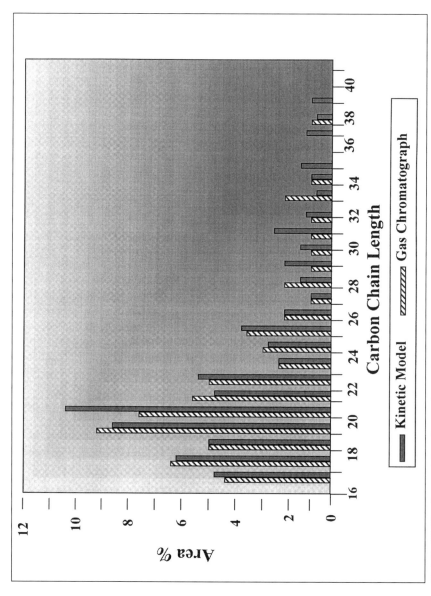

Fig. 8–10 Kinetic model and gas chromatogram

8–4. What primary difference would one expect to see between cooling rate profiles and boiling point profiles?

References

Noller, Carl R. *Textbook of Organic Chemistry.* 3rd ed. London: W. B. Saunders Co., 1966.

Becker, H. L. "Computerized Sonic Portable Laboratory." U.S. Patent No. 5,546,792, August 1996.

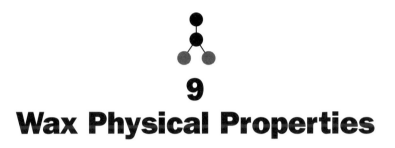

9
Wax Physical Properties

Melting Point and
Boiling Point of Alkanes

From earlier discussions the importance of alkanes in the formation of waxes present in crude oils has been emphasized. This emphasis, and the considerations of the forgoing methods of analysis, should be further discussed in terms of the bulk behavior of these structures. The following section discusses some of the physical characteristics of hydrocarbons and the effects of branching or structural isomerism (see Fig. 9–1)

When the number of carbon atoms present in an alkane increases, the number of structural isomers possible increases exponentially as shown by the following figures, which are derived by statistical calculations:

C_7—9; C_8—8; C_9—35; C_{10}—75; ...C_{15}—4347; C_{20}—366,399; C_{30}—4,111,846,763...

Boiling points

The boiling points of normal hydrocarbons increase with increasing molecular weight. These boiling points fall on a smooth curve when they are plotted against the number of carbon atoms. The increase in boiling point is due to increased attraction between molecules. Normal alkanes require around 1 Kcal per carbon atom per mole energy to convert the liquid to the vapor phase. It is not possible to distill hydrocarbons containing more than 80 carbon atoms without decomposition, no matter how perfect the vacuum, because the energy of about 80 kcal per mole required to separate the

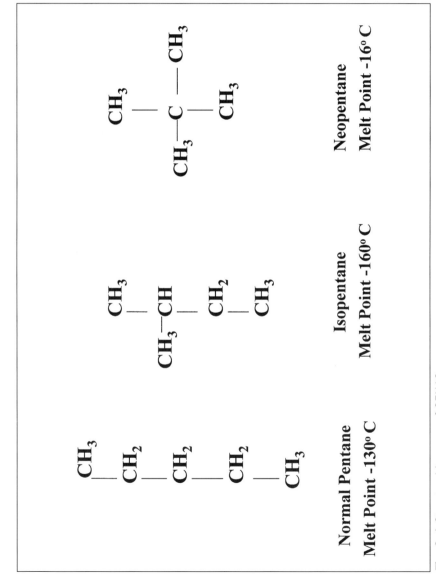

Fig. 9-1 Structural isomers of C5H12

molecules is approximately the same as that necessary to break a carbon-carbon bond.

Structural isomers or branching of the chain always result in a lowering of the boiling point. Thus n-pentane boils at 36° C, isopentane at 28° C, and neopentane at 9.5° C. Branching decreases the magnitude of the transient dipole (London forces) and prevents the attainment of optimal proximity of the molecules to each other.

Melting points

The melting points of normal alkanes do not fall on a smooth curve but show variations from expected plots. They fall on two curves, an upper one for the hydrocarbons having an even number of carbon atoms and a lower one for those with an odd number of carbon atoms.

Analysis of x-ray diagrams of crystallized solid normal alkanes shows the chains are extended, and that the carbon atoms form a zigzag arrangement. Compounds with an even number of carbon atoms have the end carbon atoms on opposite sides of the chain, while odd number chains have the end carbon atoms on the same side of the chain. The chains with an even number of carbon atoms pack more closely, making the London forces more effective, leading to a higher melting point.

Because melting points depend on how well the molecule fits into a crystal lattice, there is less regularity in melting point profiles than in boiling point profiles (see Fig. 9–2). Therefore, branching determines the effectiveness of the attractive forces between molecules in the crystal lattice. Consequently n-pentane melts at -129.7° C, isopentane at -160° C, and neopentane at -20° C. In general, however, the more symmetrical and compact the molecule, the higher its melting point.

Bulk System Properties

Throughout the previous discussions, an effort has been made to develop the theoretical foundations for the behavior of physical systems. Although the cumulative effects of these microscopic factors are responsible for the physical behavior of bulk systems, it is the macroscopic manifestations of these factors to which we are commonly exposed. Since the discussion has continuously narrowed to focus on the behavior of waxes, an elaboration on some of the behavior exhibited by these systems is in order.

Molecular Contribution to Bulk System Rheology

Rheology, or the study of fluid flow properties, is an area that includes the microscopic forces responsible for fluid behavior as well as the macroscopic manifestations presented by the bulk system. According to Karol Mysels, *Introduction to Colloid Chemistry*, "The ability to flow is, by definition, the

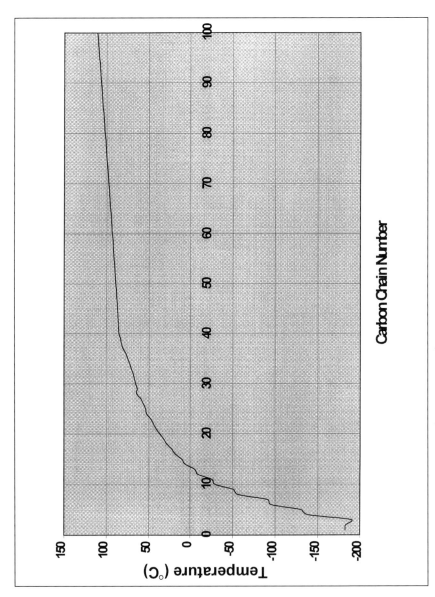

Fig. 9–2 Melting point range of normal paraffin

most general property of liquid and semisolid materials, and comes into play continuously whenever we deal with these." Further, viscosity is a quantitatively measurable subset of rheology, and is defined as "poise...the force in dynes required per square centimeter to pull two large planes, 1 cm. apart, with a relative velocity of 1 cm/sec."

$$\eta = Stress/Rate\ of\ shear = (F/A_r)(\Delta u/\Delta x),$$

where

F	is the total force acting over the area
A_r	is the area
$(\Delta u/\Delta x)$	is the velocity per unit-spacing

In the c.g.s. system, the dimensions of viscosity are dynes-sec/cm.^{2} or g/sec.-cm. The c.g.s. unit of viscosity is the poise." This definition of viscosity, then, makes it clear that the summation of microscopic interactive forces responsible for aggregation will act in opposition to applied external motive forces.

Implicit to the viscosity expression $(\Delta u/\Delta x)$, which is the summation of laminar velocities with respect to summation of the individual lamella (layer), is the summation of frequency $(1/\Delta t)$. If this implicit relationship is looked upon as simply a means of deriving the appropriate units of viscosity, or poise, by its multiplication with the quotient of F/A, its significance is lost. However, looking upon this ratio as representing a statistical relationship between the components of one lamella with those of another, a deeper understanding emerges.

The significance of this integral frequency factor $(1/\Delta t = \Delta v)$ is that it averages the resultant vector sum of the individual component velocities within each lamella and the external motive force. Another way of stating this is that the time average contribution to net fluid movement is a result of the force applied in the direction of the desired movement versus the forces acting in opposition to the desired direction of movement. This explanation explicitly indicates the importance of temperature, since molecular velocities are solely responsible for its expression.

The statistical nature of temperature is well documented, and the effects of temperature on the viscosity of fluid systems is also generally, if not quantitatively, understood (see Fig. 9–3).

An analog to the Arrhenius expression is used to explain the qualitative effects of temperature on viscosity:

$$\eta = Ae^{\Delta E(visc.)/RT}$$

where

A	is some undetermined constant

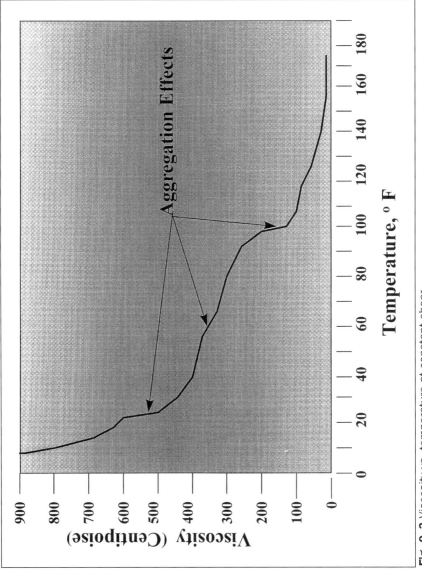

Fig. 9–3 Viscosity vs. temperature at constant shear

If the two expressions above are equated and rearranged, the statistical nature of the *(Δu/Δx)* term becomes clear.

$$(Ae^{\Delta E(visc.)/RT}/(F/A_r) = (\Delta u/\Delta x)$$

The implication of this derived relationship is that the statistical average of viscosity divided by the force acting on an area is directly proportional to the change in frequency of the interactions taking place between lamella. Thus gross viscosity effects are due to the individual quantum effects (London forces) of the molecular wax aggregates.

Quantum Considerations of Viscosity

Let's examine the diameter of the pipe as a function of the material that adheres to the pipe wall, reducing the effective diameter, and thereby reducing flow. If we look at the pipe in cross-section, and consider the center to be representative of a point of highest temperature and the outer diameter to represent the lowest, then we can describe heat radiation as a function of the diameter (see Fig. 9–4).

In order to do this we can use the Stefan-Boltzmann relations for radiant heat from a cavity and a surface as a function of temperature.

$$Rc = \sigma T^4 \ (Cavity)$$

and

$$R = \epsilon \sigma T^4 \ (Surface)$$

where

σs	is a universal constant (the Stefan-Boltzmann constant) whose measured value is 5.67×10^8 watt/(meter^2)($^{\circ}$ K^4)
ϵ	is the emissivity and depends on the material and the temperature

Thus, if we know the diameter of the pipe, and the temperature at its center, we can calculate the power (watts) radiated from its center. If we know the emissivity of the pipe, and we measure the power (joule/seconds) emitted at its surface, we can determine the summation of the intervening (center to pipe surface) emissivities. The significance of this relationship is illustrated by the fact that the integral (summation) of the wavelengths of the cavity radiance is independent of the material and the shape and size of the cavity.

$$R = \int R_\lambda d\lambda$$

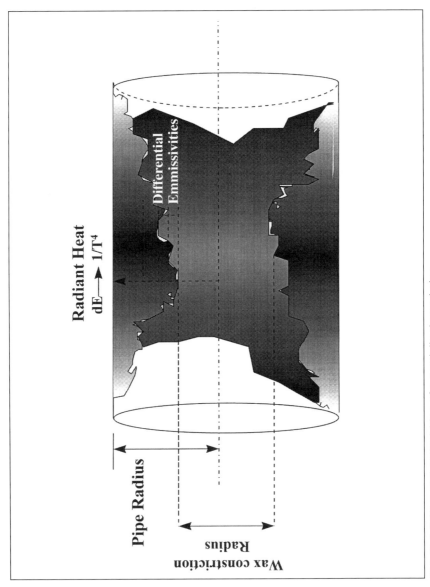

Fig. 9–4 Differential emissivities of a partially clogged pipe

where

$d\lambda$ is the change in radiant wavelength

R_λ is spectral radiancy, defined so that the quantity $R_\lambda d\lambda$ is the rate at which energy is radiated per unit area of surface for wavelengths lying in the interval $\lambda + d\lambda$ (watts/cm^2)

This information, combined with the following equality leads a significant conclusion:

$$R_\lambda = 2\pi c^2 h / \lambda^5 [1/(e^{h\upsilon/kT}-1)]$$

where

c is the velocity of light in vacuum

h is Planck's constant (6.63×10^{-34} joule-seconds),

λ is the radiant wavelength

k is Boltzmanns' constant [1.38×10^{23} joule/ (molecule)($^\circ$K)]

T is the temperature ($^\circ$K)

$\upsilon = c/\lambda$ is the frequency of radiation.

This formula constitutes one of the most important formulas derived in physics, because it introduces the concept of quantum mechanics for which Max Planck received the Nobel Prize. However, it should be noted that, since $R_\lambda d\lambda$ represents the rate of energy radiation per unit of surface area (kg/sec^3), this quantity is time dependent. The result of all this is that the individual molecular interactions and their number give rise to energy that is emitted as radiant, and the quantity is measurable as blackbody radiation (see Fig. 9–5).

What is blackbody radiation? If a quantity of gas is confined to a box where the walls are maintained at a constant temperature, the average kinetic energy approaches a limit per molecule, and average kinetic energy distribution among the molecules takes a definite form. In the relaxed or final state of the gas, the average kinetic energy per molecule is a fixed multiple of the temperature, and the distribution of kinetic energies is described by a universal function of the ratio K/θ between the kinetic energy K and the temperature θ. If the walls of the box absorb and reemit radiation, it will eventually be filled with radiation that is equally intense in all directions. The distribution of photon energies is then described by a universal function of the ratio of $h\upsilon/\theta$, the equilibrium value of which is called *blackbody radiation*.

The consequences of these quantum effects, although seemingly remote to the discussion of bulk fluid behavior, are numerous and profound. The forces leading to aggregation and the applied external forces act in opposition to each other, and depending on which force is superior, the fluid will either flow, or it will not. When the aggregation forces are superior, the

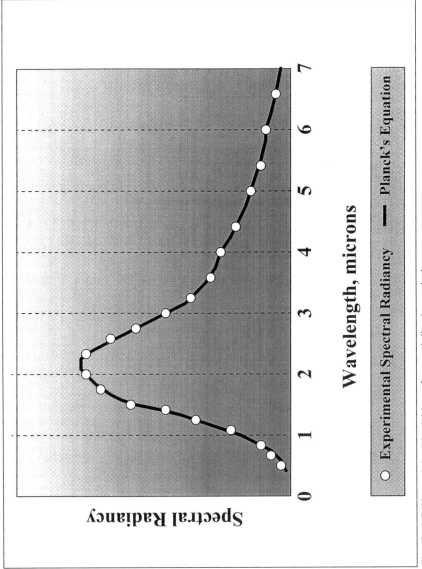

Fig. 9–4 Differential emissivities of a partially clogged pipe

intermolecular forces of attraction produce aggregates that can continue to interact with other molecular aggregates, or they can add to surface imperfections on the internal pipe surface. This interaction with the pipe surface and subsequent aggregation with other molecular surfaces results in a reduction of the pipe's diameter.

The increased turbulence is caused by the fluid velocity increases, which result from greater applied pressure per unit of unobstructed cross sectional area, and act to shear loose those aggregates that are softer, leaving behind only the strongest aggregates. Thus, as time goes by the pipe diameter decreases, and the deposits become harder until finally the flow ceases. This gross manifestation of fluid behavior should be considered the result of the summation of the microscopic forces.

Thus, when aggregates form and cause constrictions within pipes, the insulative effects (radiant heat absorption) of the constriction should also be considered. These insulative effects reduce the capacity of the fluids to transfer energy via radiation, and therefore magnify the problem. Another way of stating this is that these insulative effects result from the change in the system's ability to transmit or adsorb quanta (hv).

Intrinsic Viscosity

Intrinsic viscosity is defined as the nonzero viscosity intercept extrapolated to zero concentration of solute, which is a measure of the solute/solvent interaction. The intrinsic viscosity is generally designated by the symbol η. Since this quantity is taken to represent a minimum as the concentration of solute approaches zero, it is assumed that the configuration of the solute is in a relaxed condition, and occupies a preferred, or optimal, solvent accommodation. An expression known as the Mark-Howink relation can be used to determine values related to the shape and size of the particles giving rise to the intrinsic viscosity:

$$[\eta] = K\,M$$

where

K is a proportionality constant whose value depends upon the solvated configuration of the solute, an exponential term that depends on the solid geometry of the solute

M is the particle weight

A consequence of this definition for intrinsic viscosity is that high molecular weight polymers can be solvated in dilute systems, and a fairly accurate measurement of the molecular weight can be made. This relationship also suggests that systems containing solvated high molecular weight polymers exhibit higher viscosity. Thus, an operational use of the term intrinsic viscosity has been taken to mean that systems containing high molecular weight polymers exhibit intrinsic viscosity effects (see Fig. 9–6).

Pseudoplasticity and Thixotropy

According to Karol Mysels, "If the particles form aggregates and the aggregates can be broken up by shear, their size will depend on the rate of shear." This phenomenon is called thixotropy and is a subset of the class of viscosity behavior known as pseudoplasticity. Another behavior exhibited by fluid systems is called dilatancy. If particles experience increased encounters as shearing forces are applied, and these encounters result in aggregations of the particles, then shear increases the viscosity, and this phenomenon is called dilatancy. The dilatant behavior is very seldom seen, but thixotropy is commonly observed. Another name for pseudoplasticity is non-Newtonian fluid behavior.

Newtonian fluid behavior is exhibited by systems of fluids that undergo no aggregation or disaggregation as a result of shear forces. Waxes represent a class of materials that display thixotropy. Depending on the composition and concentration of the waxes present, these systems will give a large hysteresis (shear stress versus shear rate energy envelope resulting from starting at low shear stress proceeding to high and returning to low; see Fig. 9–7 and Fig. 9–8).

Yield Value

From the graph showing the phenomenon of hysteresis, it can be seen that at the shear stress intercept there are two points where the curve crosses. The first point, (usually the highest point) is called the up yield value, and the second is the down yield value (see Fig. 9–8). These values represent the shear force necessary to overcome the aggregation forces of the components before and after undergoing a shear force respectively. Thus, these measurements are intended to reflect the behavior of fluids in a flowing system before, during, and after an application of shear.

Yield measurements are usually conducted under conditions of constant temperature and constant pressure to avoid the obvious complications of changing aggregate arrangements. Thus, a family of hysteresis loops are obtained by the performance of a series of yield-point tests at different temperatures and pressures.

An implicit test methodology is then suggested from the yield-point test's sensitivity to temperature and pressure changes. This test methodology involves the application of a constant shear over changing conditions of temperature, pressure, or both. In thixotropic systems, such as those containing waxes, the viscosity is observed to increase exponentially as the temperature is lowered. Thus, curves generated in this way produce characteristic traces of increasing viscosity versus decreasing temperature at constant shear stress. The family of curves generated in this fashion shows several features about the sample under study.

One of these features is the result of aggregation of components as the temperature is lowered, which results in an abrupt deflection in the otherwise smooth changing slope of the curve. From the description of Newtonian fluid

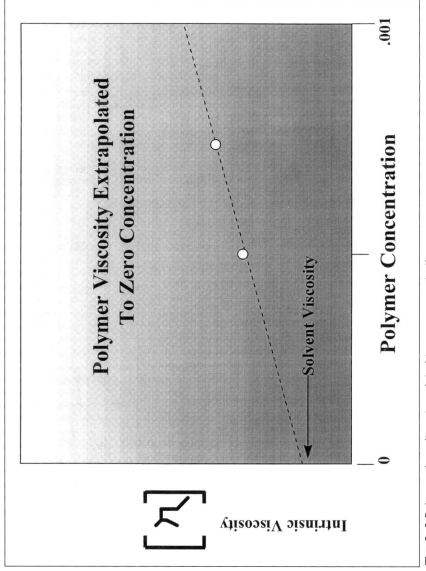

Fig. 9–6 Polymer viscosity extrapolated to zero concentration

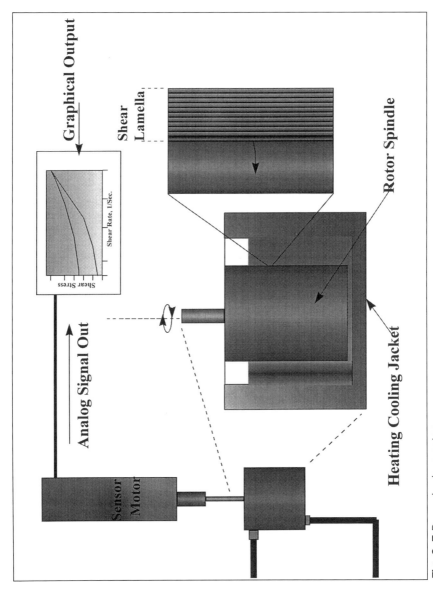

Fig. 9–7 Dynamic viscometer

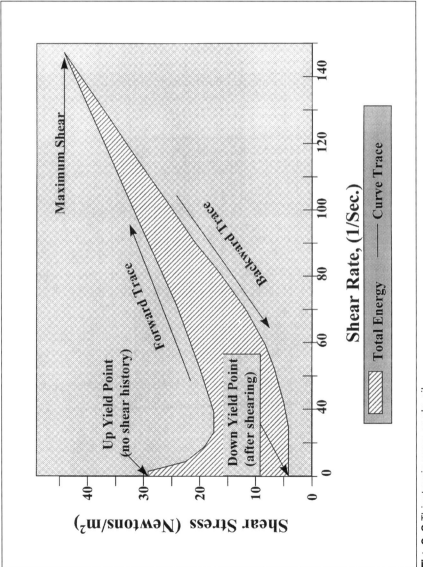

Fig. 9–8 Thixotropic waxy crude oil

mechanics above, the curve of a Newtonian fluid would be expected to exhibit a smooth exponential curve under these conditions. Instead, pseudoplastic fluid systems show deviations from smoothness that result from component aggregation under changing temperature or pressure conditions. It is these changes that are interpreted by the technician as cloud point (that temperature at which the first deviation from curve smoothness is observed) or primary aggregation, and subsequent aggregation is secondary, tertiary, etc.

Analogous Thermometry and Rheology

Dynamic rheology, as discussed above, is a mechanical method of measuring the effects of aggregation on bulk fluid behavior, and as such is subject to limitations of instrument sensitivity. Likewise, thermometric measurements are subject to the same limitations of sensitivity. However, the magnitude of mechanical forces required to alter the environment of the component aggregates leading to a measure of their resistance to the applied shear forces is large. Consequently, those components representing the onset of aggregation are swamped by the application of overwhelming shear forces.

Thermometric measurements attempt to measure the energy released by the aggregating components and are generally conducted with a minimum of externally applied shear forces. However, since the measurement of temperature is dependent upon the populations of components involved in aggregation (e.g., within a velocity domain) the energy evolved as heat is small for low populations. Thus, both methods are subject to population effects, and therefore, represent a statistically dependent measurement.

The implications of this statistical dependence are that one, the other, or both methodologies can be improved by increasing the number of observations per unit time, and/or expanding the duration of conditions under which the measurements are taken. Implicit in each or both of these options is the attainment of equilibrium. Practical limits of observational time periods are a part of life, and as such render the extension of these periods impractical. Thus, a compromise is called for, and the employment of increased numbers of measurements over some reasonable extension of the observational period is appropriate.

An additional complication arises, however, when the number of observations per time period is increased—this complication is the phenomenon of noise. Noise is the result of spurious signals that arise in the electronics of measuring devices. These signals have many causes, but their importance is that they superimpose themselves on the signals from the sample under study. The randomness of these noise signals provides a means of their elimination. Fortunately several filtering devices and algorithms are available to accomplish just this task.

Sound As a Means of Measuring Aggregate Behavior

Sound represents a wave phenomenon that results from minute displacements of the equilibrium positions of those components that comprise the media through which it is propagated. Sound has a source, and as such an initial positional displacement with respect to time, and therefore an initial velocity. The velocity with which it is propagated through a given media is a directly proportional function of the density of the media. Thus, the separations between the components and the mass of the individual components comprising the propagation media determine the velocity of the sound wave.

Inherent to this description of the propagation media is its uniformity of composition. If the propagating media is nonuniform, for example a solid containing pockets of air, then the time period required for a sound wave to traverse a given distance will be slower than expected through the uniform solid. Analogously, sound produced in one room passes through a wall and is received in the opposing room; the time taken for the sound to travel from its origin to its destination is shorter than it would have been if there was no intervening wall. This characteristic of sound propagation then can be used to an advantage when systems of solids, liquids, gases, and combinations of thereof are studied.

Accompanying the above behavior of sound propagation through a media are secondary effects that can be employed profitably when physical systems are studied. One of these effects is the echo effect. This effect is possible because sound waves are incapable of superposition, since they constitute no physically unique positional displacement. This means that sound waves, irrespective of their origin, can occupy the same x, y, and z coordinates, but they must be temporally separate (they cannot occupy the same time coordinate).

The result of this *nonsuperposition* is that a sound wave undergoes constructive (strengthening) and destructive (weakening) interferences that do not change its propagation velocity (since the composition of the media remains constant) but do act to change its amplitude and wavelength. These changes in amplitude and wavelength can then be measured and examined to gain some insight into the nature of the propagating media.

If the above scenario of analysis is conducted under constant conditions of temperature and pressure, information about the propagation properties of the media are obtained that are specific to those conditions. However if the conditions are changed, then the changing characteristics of the media's propagation properties are obtainable. Herein lies the potential for the examination of aggregate phenomena (see Fig. 9–9).

Even when sound is used, we are plagued by the statistical nature of aggregate phenomena, since the size of the perturbation imparted to the wave is a result of the population of interacting components within a velocity range. Thus the population of components occupying equilibrium velocity envelopes that are disturbed by the impression of a sound pressure is directly proportional to the contribution they make to the wave distortion.

There is, however, an inherent advantage to the use of sound for these measurements. The advantage is that the range of velocities exhibited by the interacting components is of the same order as that of the velocity of the

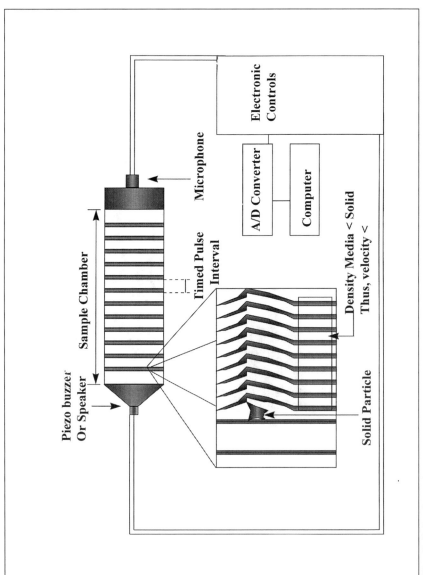

Fig. 9–9 Sonic probe system

sound's propagation. Consequently, harmonic (optimizations of amplitude) effects are experienced, which enhance the accuracy of the measurements. Additionally, vastly increased sampling is possible because of the continuous nature of applied force and the speed of its propagation.

Enhancements to the utility of this methodology include extremely short sound bursts (velocities are measured by crystal oscillators) and combinations of crystal oscillators, which measure the time of travel of the outgoing and echoed or incoming resultant wave. By these methods it then becomes feasible to echo-locate aggregates, and possibly display their time-averaged location as an image of the confined space containing the media being examined.

Summary

The preceding chapter is a discussion of some of the techniques employed for the characterization of bulk fluid behavior and some of the instrumentation used. This discussion focuses on the heavily used dynamic viscometer, and how some of the results using this instrument can be related to individual fluid component contributions to the bulk fluid's behavior.

The sonic test method was also discussed and how it might be employed to better characterize fluid systems. An effort was made to tie together some of the macro phenomena exhibited by bulk systems and the micro phenomena that act in concert to produce the bulk behavior. An introduction to descriptive rheology was among the system properties discussed. Viscosity, and the interactions responsible for its different manifestations, were examined from both mechanical and quantum mechanical points of view. Heat radiation was examined from the differential emissivities of the contributing structures, and a tie-in to the phenomena known as blackbody radiation was developed.

Problems

9–1. What effect on melting point would one expect to see as a result of branching?

9–2. In a plot of viscosity versus temperature, draw curves showing inflections for aggregate interaction effects expected in the following systems: butter, peanut butter, and molasses.

9–3. Give a qualitative description the radiant heat effects expected at the surface of a pipe inside which accumulations of wax are forming.

9–4. Propose a detection device to measure the radiant energy at the surface of a pipe.

9–5. Give a physical description of thixotropy and dilatancy.

9–6. Describe in words and diagram form the meaning of first (up) yield point, and second (down) yield point.

9–7. Give a rheological definition of a crude oil cloud point, and describe how it relates to aggregate effects taking place in the oil.

9–8.   Give a description of how the superposition principle applies to sound waves.

9–9.   What differences do wave propagation which involve displacement from molecular equilibrium position and self propagating waves exhibit with respect to the superposition principle?

9–10. What forms of energy transfer are involved in heat release, or uptake in aggregate systems? (Extra credit)

References

Barrow, Gordon M. *Physical Chemistry.* 2d. ed. New York: McGraw-Hill Book Co., 1966.

Becker, H.L. "Computerized Sonic Portable Testing Laboratory." U.S. Patent 5,546,792, August 20, 1996.

Layzer, David. *Constructing the Universe.* 1st ed. New York: Scientific American Library, 1984.

Mysels, Karol J. *Introduction to Colloid Chemistry.* 1st ed. New York: Interscience Publishers, Inc., 1959.

Resnick, R., Halliday, D. *Physics, Part I.* 1st ed. New York: John Wiley & Sons, Inc., 1966.

10
Wax and Quantum Effects

Electromagnetic Effects of Aggregation

In the section "Quantum Considerations of Viscosity," the Stefan-Boltzmann expression for spectral radiancy, R, was discussed in terms of the physical dimensions of a pipe, but the significance of the emissivity term, ϵ, was not adequately demonstrated. Thus, focusing attention on this term and what it represents is key to understanding the electromagnetic effects of aggregation phenomena. Since emissivity is characteristic to the material giving rise to its value, variances in the material's composition would, therefore, be expected to change its value.

The Handbook of Chemistry and Physics gives the following definition:

> *Emissive power or emissivity is measured by the energy radiated from unit area of a surface in unit time for unit temperature between the surface in question and surrounding bodies. For the cgs system the emissive power is given ergs per second per square centimeter with radiating surface at $1°$· absolute and the surroundings at absolute zero.*

The defining equation accompanying this definition is given as follows:

$$\epsilon = M/M_{blackbody}$$

where

M and M$_{blackbody}$ are respectively the "radiant exitance" of measured specimen and that of a blackbody at the same temperature as the specimen.

The phrase "radiant exitance" is better understood as radiant flux density at a surface and is expressed mathematically as follows (note the similarity to the expression for viscosity):

$$M = d\phi/dA_r$$

where

ϕ	is dQ/dt
Q	is radiant energy
A_r	is surface area

Flux, ϕ, can be broken into four contributions as follows:

ϕi = incident flux
ϕa = absorbed flux
ϕr = reflected flux
ϕt = transmitted flux

But we know that Planck's expression for radiant energy is $h\upsilon$, therefore, $Q = h\upsilon$. If $Q = h\upsilon$, then $dQ/dt = h\ (d\upsilon/dt) = h\ (d(c/(\lambda)/dt) = hc\ (d(1/\lambda)/dt)$, which upon integration gives the following:

$$Q = hc\ [1/\Delta\lambda]$$

Thus, $d\Delta = d^2Q/dt^2 = hc\ d^2(1/\Delta\lambda)/dt^2 = h\ d^2\Delta\upsilon/dt^2 = h\ d^2(1/\Delta t)/dt^2 \rightarrow h/t^3$, which finally results in the following expression:

$$M = d\phi/dA_r = h/(t^3 dA_r)$$

which, when dimensionally analyzed, yields the units of mass/(time)^4 or kg/sec^4.

So that the emissivity factor Œ is a ratio of mass given up by radiant energy evolution as a function of quadric time. This result differs by ΔT/*sec*.** from the definition of emissivity given above $(T_{surroundings}-T_{surface})(ergs/[(sec)(cm^2)]$, which yields $(\Delta T)(mass)/sec^3$. It should also be noted, from the definition above, that the term represented by ΔT is negative, since the radiating surface is at 1° K, and that of the surroundings is 0° K. Thus, the difference between the two results can be rectified by multiplying the second derivation by $i2T$, which could also be expressed as it (from the complex plane of time), where $i = (-1)^{1/2}$, and t is time. (Note: Here again, the discussion of the relation $iT = t$ and the units of the ratio (h/nk) in chapter 4 can be helpful in clearing up the apparent anomaly.)

After having gone through the above exercise, it is possible to develop the differential form of the emissivity values resulting from aggregate interaction. If we realize that the differential form of the emissivity gradient represents the individual component's contribution to the observed emissivity, then we see that the integral is just the emissivity. Or the emissivity is $\int d\epsilon_s$, and the individual emissivities of the component emissivities are represented by ϵ_s.

Considering this result, the fact that this energy is emitted as radiation, and that radiation is an electromagnetic wave, the possibility of employing the superposition principle for analysis becomes evident. According to Kenneth Ford's *Basic Physics*,

> *The vectorial combination of differently polarized waves illustrates one aspect of the fundamental idea of superposition. . . .the superposition of electric or magnetic fields means that the field at a point arising from several sources is the vector sum of the separate fields that would be present if each source were acting alone. Physically, it means that fields coexist in space without influencing each other. If a second field is added where a field exists already, the first is unchanged by the second, and the second is uninfluenced by the first. The total field is the sum of the two.*

Thus the alterations of these waves through either constructive or destructive interferences is not expected. Therefore, the absence of these effects suggests that diffraction studies on aggregating systems could yield useful information about the systems.

Some microscopic studies have been conducted employing lasers (see Fig. 10–1). The sample under study is placed on a microscope slide, which can be heated or cooled by circulating variable temperature fluids on the underside of the slide. The laser light source is fixed at an angle of approximately $60°$ to the objective lens of the microscope, or $60°$ to the normal of the slide. The sample is heated to a temperature above the melting point of the solid waxes visible at a magnification of 100 x. The temperature is then gradually dropped, and the magnified image recorded on video tape.

At higher temperatures the laser light is completely transmitted, but as the temperature is lowered, solid aggregates are observed by defracted laser light. It is because of this behavior, and the above discussion of emissivity, that a methodology employing monochromatic radiation at precise angles affords an opportunity for spectral analysis of aggregate phenomena.

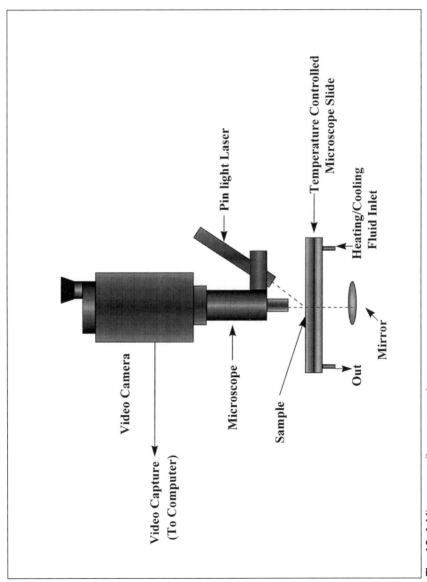

Fig. 10–1 Microscope/laser apparatus

Waxes and Piezoelectricity

A very interesting phenomenon known as the piezoelectric effect is exhibited by certain types of crystals that (according to the *Handbook of Chemistry and Physics*), "...is only possible in crystal classes [that] do not possess a center of symmetry." Further, this effect is defined as "the phenomenon exhibited by certain crystals of expansion along one axis and contraction along another when subjected to an electric field. Conversely compression of certain crystals generate an electrostatic voltage across the crystal."

An interesting combination of questions now arises. Do paraffin wax crystals have a center of symmetry? Is lack of symmetry a necessary and sufficient condition for piezoelectricity? If we accept the London explanation of intermolecular charge induction forces, the answer to the first question is no, since along any axis of two interacting paraffin molecules the charge centers are of opposite sign.

The answer to the second question is somewhat more complicated, and can only be suggested by inference. According to the *Handbook of Chemistry and Physics*:

> *Certain waxes, when polarized in their molten state, retain a permanent polarization after solidifying, even though the external polarizing field is removed. Electrets, manufactured in this way, are the electrostatic analog of permanent magnets in that they possess a gross permanent electric dipole moment. Materials from which electrets can be constructed are called ferroelectric.*

The degree to which paraffin waxes are polarizable is determined by their dielectric constants (2.0-2.5 x 10^6 cycles) and the electric field strength (250 volts/mil). Thus, the inferred answer to the second question is no, since polarizability and conductivity must also be possible for piezoelectric effects to be exhibited.

Given the above discussion, paraffin waxes would be expected to exhibit piezoelectric effects provided the crystals can carry the charge produced by their deformation. Thus, the ability of a paraffin wax to conduct charge will be a function of its conductivity or resistivity. Paraffins exhibit very high resistivity $\sim 10^{15}$–10^{19} ohms/cm at $23°$ C, and consequently the voltage required to cause polarization and thus give rise to piezoelectricity would also be expected to be high.

An additional factor determining whether paraffin waxes exhibit the piezoelectric effect or not is the strength of the crystal lattice. If the crystal lattice forces are weak, then the deformation of the crystal from its equilibrium value may be insufficient to cause a voltage high enough to be conducted by the wax. Since the forces holding wax crystals together are weak (London/van der Waals), and the conductivity is low at or near the melting point of the wax, normal distortion forces would not be expected to generate piezoelectricity.

However, since waxes continue to contract with decreasing temperature, the lattice forces increase in strength, so that the difference between distortion force and crystal forces narrow. Although the conductivity also decreases with decreasing temperature, it is reasonable to assume that at a specific temperature the piezoelectric effect will be observed. If this piezoelectric effect is exhibited, then the temperature at which it occurs should correspond to a critical point. This critical point temperature could then be considered the superconductivity temperature, where resistivity is negligible and conductivity extremely high.

Another interesting and important phenomenon exhibited by certain crystals is that of sympathetic vibration. A good example of this is found in the quartz crystal oscillator. A piece of quartz crystal sandwiched between metallic plates deforms mechanically when a potential is applied between the plates. When the external potential is removed, the relaxation of the crystal will induce a potential between the metal plates (piezoelectricity). There will be a natural vibration frequency of the crystal that depends on its cut and size. An A/C voltage of the resonant frequency of the crystal applied across it will maintain the resonant vibration (see Fig. 10–2).

Temperature has a pronounced effect on the oscillation frequency of the crystal and acts to increase the amplitude as it increases and decrease the amplitude as it decreases in accordance with the accompanying changes in resistivity. Here again population effects become pronounced as indicated by the curves in plot of frequency versus amplitude. It is interesting to note that the piezocrystal imposes its own vibrational frequency on the ICR (inductor, capacitor, and resistor) circuit, and since the following relationships are true at resonance, some additional extrapolations can be made:

$$\omega" = (1/LC)^{1/2}$$

$$I_m = \xi_m/R \text{ (at resonance)}$$

$$\omega" = \omega = 2\pi\upsilon = (k_m/m)^{1/2} = (1/LC)^{1/2} \text{ (at resonance)}$$

where

I_m	is the current amplitude
L	is the inductance
C	is the capacitance
R	is the resistance
$\omega"$	is the angular frequency
ω	is the undamped angular frequency
υ	is the frequency of oscillation
ξ_m	is the variable voltage (A/C)

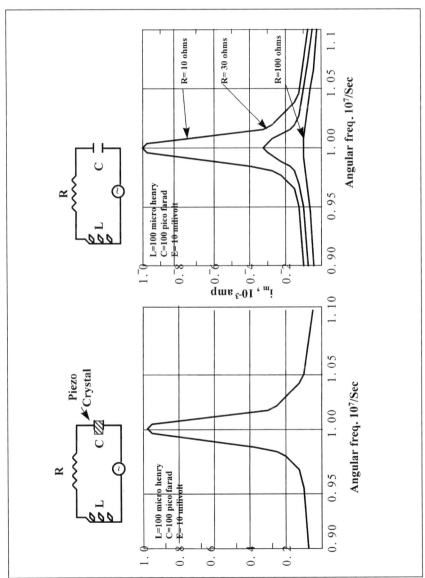

Fig. 10-2 Angular frequency is dependent on piezoelectric crystal frequency when an alternating current

Combining these relations with the following allows us to draw some conclusions about the nature of the crystal time constant and temperature.

$$I = \xi/R \; (1-e^{t/\tau})$$

$$\tau = L/R$$

where

τ is the inductive time constant (dimensions = sec)

If L/R is set equal to 1, then the significance of the time constant shows the following relationship:

$$I = \xi/R \; (1-e^{1}) = (1 - .37) \; \xi/R = 0.63 \; \xi/R$$

Thus the time constant t is that time at which the current in the circuit will reach a value within $1/e$ (about 37%) of its final equilibrium value. Thus, if I is converted to Im the following relationship is derived:

$$I_m = (+/-) \; \xi/R \; (1 - e^{1}) = (+/-) \; 0.63 \; \xi/R$$

where

$(+/-) \; \xi$ is equal to ξ_m

This gives us the wave form sinusoidal character.

If v is the harmonic frequency of the crystal, then at resonance, v will determine τ or L/R.

$$R = \rho \; (l/A_r)$$

$$\rho = \rho_o \; (1 + \alpha'(T - T_o))$$

$$\alpha' = (1/\rho_o) \; ((\rho - \rho_o)/(T - T_o))$$

where

R	is the resistance (ohms)
T	is = oK > To = 0o K
ρ	is resistivity
ρ_o	is the resistivity at 0o K
l	is the length of a cylindrical conductor
A_r	is the cross-sectional area of conductor
α	is the coefficient of resistivity at T
α'	is the mean coefficient of resistivity over $(T - T_o)$

Combining this relationship with those above, the relationship of temperature to the time constant becomes apparent since $\tau = L/R = L/[\rho(l/A_r)]$. Thus, the time constant is inversely proportional to the temperature difference across a conductor of length l and area A_r or:

$$\tau = k_\rho \, (x/\Delta T)$$

where

k_ρ	is the constant of proportionality (, K sec/m)
x	is $(A_r \, /l)$

Now it is possible to derive the relationship between spectral radiancy (R) and the time constant τ.

$$R = \sigma \epsilon T^4$$

$$\Delta R = \sigma \epsilon (\Delta T)^4 \text{ (differential Spectral Radiancy)}$$

$$\Delta R = \sigma \epsilon (k_\rho x/\tau)^4$$

This can finally be stated: "the spectral radiancy changes as the fourth power of velocity."

Practical Applications of Electromagnetic Aggregation Effects

The treatment of the quantum mechanical and electrodynamic concepts involved in aggregation phenomena was presented to develop a deeper understanding of the microscopic picture of crystal formation. It is through this deeper understanding that development advances are made. At this point in the discussion it might be helpful to try to describe some of these concepts in more common fashion, or one that fits our everyday experiences.

When a reference to spectral radiancy R is made, it should be clear that this represents an emission of energy by radiation (electromagnetic waves) from a given object of unit surface area to the surroundings. Thus, when any object cools, radiation is emitted from the object of highest temperature to an area of lower temperature. This radiation can be conducted via self-propagating electromagnetic waves without the necessity of an intervening material transport media.

If a wire is heated to glowing in a perfect vacuum, a thermometer placed within the vacuum envelope will not register a change in temperature. However, if the thermometer is placed on the surface of the containing vacuum envelope (glass surface of light bulb), under atmospheric conditions, the thermometer will register a temperature increase. Thus, the transfer of energy requires no transporting media, but the measurement of the energy imparted by the radiation requires a material media to convert energy to the physical manifestation of heat (see Fig. 10–3).

The transfer of energy via radiation (electromagnetic waves) does not need a transfer media, but it often encounters matter along its travels. This is when the electrons of the material through which these waves travel capture a portion of the energy they possess, and transform it into additional angular momentum. The additional angular momentum now carried by the electron places it in a higher energy level, or increased radius from the nucleus. This increased energy state is an unstable condition and the electron must now reemit a photon to again reach equilibrium orbit about the nucleus.

In the process of absorbing the energy ($h\nu$ or quanta) from the incoming wave, only a fraction is transformed to additional angular momentum; the rest is defracted. Thus, when the electron reemits a photon it is of a longer (less energetic) wavelength than the incoming wave. This procedure is operative in the transfer of heat from forming crystals to their surroundings, as is the transfer of molecular kinetic energy when molecules collide.

Molecular Collisions

Molecular collisions take place with velocities that are near that of the speed of sound. Because this speed would be expected to add very small momentum contribution to a single molecule, there must be a large number of collisions taking place to produce an appreciable temperature effect. However, as a system containing wax is cooled, the temperature is seen to change less rapidly, suggesting that momentum contributions to temperature changes are also taking place more slowly.

A measurement of temperature is a measurement of the collective bulk system's physical state, not an individual aspect or component. Therefore, when a plateau of temperature versus time is seen, it is the result of the crystal's excess momentum being transferred to the fluid fraction of the solute (wax)/solvent (oil) mixture. The forming wax crystal increases in mass by increments of the molecular weight of the molecules that are combining, and consequently its velocity is decreased dramatically with each increasing increment (this is required by the law of conservation of momentum). If there were no crystals forming, there would be a smooth decline of temperature.

The discussion of energy transfer could, in some respects, be easier to understand if instead of thinking of temperature as a variable, we changed time to a variable. If temperature is considered to be the determinant of the passage of time, then we could view the crystallization process as taking place in opposition to the forces responsible for the tendency of systems to seek a state of maximum disorder.

Since this view emerges as a result of the third law of thermodynamics, reaching a temperature of absolute zero is not possible, since perfect order does not exist. If we look at the temperature-time plateau again, we could choose to think that the ordering processes of crystallization expand that

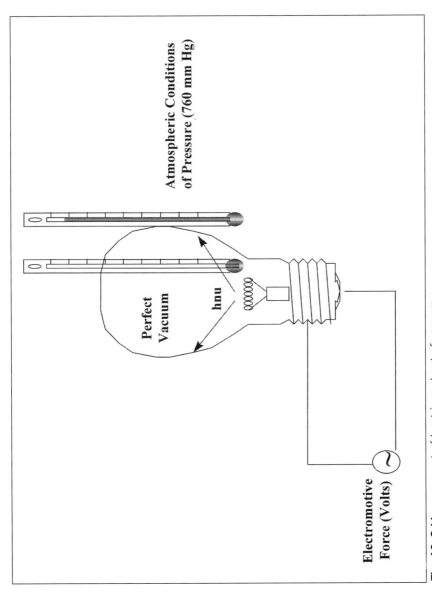

Fig. 10-3 Measurement of heat in and out of vacuum

time. This is only a little more satisfying than the opposite point of view, but it does suggest that the crystal is bucking the natural trend to a state of maximum chaos by refusing to yield to the passage of time.

Taking Advantage of Crystallization

Once we have determined the forces involved in the crystallization process, we can begin to think how we can take advantage of some the processes to alter the physical state of the bulk system. Up to this point the types of interactions taking place to form crystals have involved the process of transferring energy, either through mechanical or electromechanical means. The reason for this is quite simple; there are just no other obvious chemical handles. So we have to content ourselves with this information, and hope it can be put to use to alter the character of problem waxes.

Early on a discussion of crystal modifier interactions was conducted, and the concept was to build a molecule that would co-crystallize with the waxes to change their morphology. This method is time-honored and represents an effective technology. But from the forgoing discussions, might not some other mechanism of interference with the normal crystallization process be feasible? If a sympathetic vibration centered at the fundamental frequency of forming crystal could be set up, it might be possible to alter the morphology of the wax. Another method might be to add a component that could act as a vibrational sink, adsorbing energy from the environment and supplying it to the prospective wax crystal.

How Crystal Modifiers Work

At this point we will try to bring together some of the concepts presented in previous sections to act as aids in understanding how crystal modification takes place. Throughout the previous discussions a considerable emphasis has been placed on the importance of energy state functions. These functions are generally described in terms the contributions of the various forms of molecular energies as follows:

$$E - E_0 = E_{trans} + E_{rot} + E_{vib} + E_{elec} = 3/2\ RT + 3/2\ RT(or\ RT) + E_{vib} + 0$$

where

R	is 1.987 cal/mole °K
E_{trans}	is translational energy
E_{rot}	is rotational energy
E_{vib}	is vibrational energy
E_{elec}	is electronic energy

Table 10–1 shows the molecular energy level spacing of these energy contributions to the total molecular energy.

Table 10–1

Molecular energy level spacing (Source: Gordon Barrow, *Physical Chemistry*, 1966)

	Typical energy-level spacing ergs/molecule	cal/mole	Thermal energy E- E_0 cal/mole
Electronic	10^{-11}	100,000	0
Vibrational	10^{-13}	1,000	$RT(e^x-1)$
Rorational	10^{-16}	1	3/2RT (linear), RT (non-linear)
Translational	10^{-34}*	10^{-18}	3/2 RT

*Note how small the translational energy level separations are and that this results in confusion about their quantum nature

Considering the extremely small energy level spacing of translational energies and the comparable contributions they make to the total molecular energy, it is easy to see why the quantum mechanical descriptions of these forces blend with the classical mechanical ones.

The energy level spacing of rotational energy can also be expressed using classical mechanics; however the vibrational and electronic require the discipline of quantum mechanical constructs (see Fig. 10–4). Because the energy level spacing is small in the cases of translational and rotational energy, the contributions of the individual molecules to the total energy is also small.

This fact shows that if at a particular temperature the cooling rate is slowed, it is due to a very large concentration of interacting molecules emitting energy to the surrounding system. Given the thermal energy contributions outlined above, this transfer of energy may occur by interconversions of molecular energy (e.g., rotational to translational, translational to vibrational, and vice versa). (Note that electronic contributions are also possible, but under ordinary temperature conditions these contributions would not be expected to be appreciable.) What are some of the things that happen to these forces of interaction when a synthetically derived material is introduced into a system containing wax?

Looking at a polymer (Fig. 10–5) containing several repeating subunits with molecular weights approximating those of the waxes present, it is plain that this molecule will also have a considerably lower average velocity. The conservation of momentum demands that as an inertial mass increases, its velocity must decrease to maintain the same momentum. Thus, the polymer maintains a smaller average velocity than the wax molecules.

It should also be noted that as the polymeric modifier co-crystallizes with the incoming wax, its mass is also increasing, and therefore its velocity is slowing. This reduction in velocity results in the same transfer of translational momentum by the wax/polymer to the mix, since momentum of the system must be conserved. But the summation of velocities in the system is less

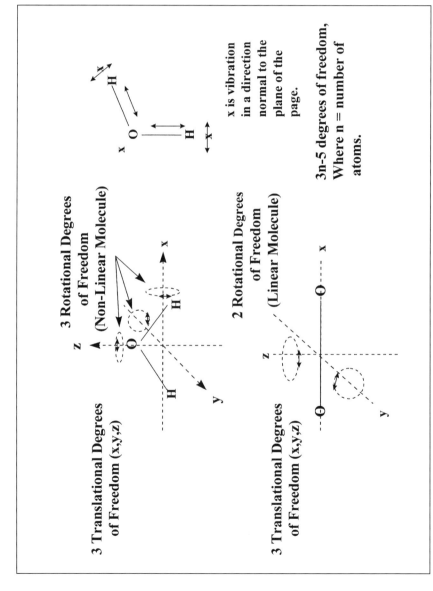

Fig. 10–4 Rotational degrees of freedom, linear and nonlinear molecules

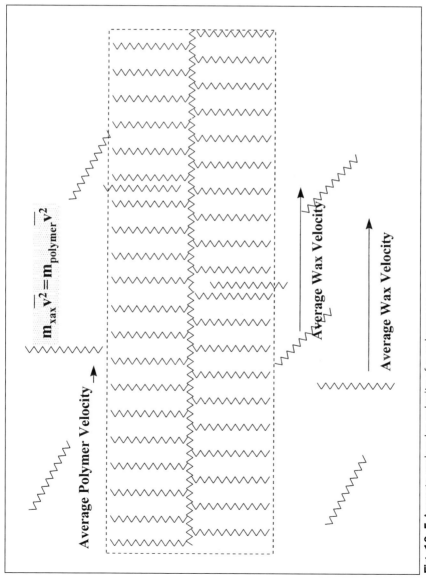

Fig. 10-5 Average molecular velocity of a polymer vs. wax

than it was before the wax's combination with the polymer, and the temperature of the system must drop (Maxwell's speed distribution).

The implications of this type of interaction and its resultant change in the transfer of energy are that a waxy oil exhibiting a temperature plateau over a cooling time must do so in a similar fashion. Thus, these plateaus are significant landmarks that should be considered as representing structures that are of key importance when attempts are made to produce crystal modifications.

Complex mixtures of wax in oil are not monocomponent waxes. They represent a range of carbon number (chain length), and these various components co-crystallize as the temperature drops. But in many cases there are specific wax components that lead to the formation of secondary, tertiary, quaternary, and higher wax aggregates. Thus if this component fraction can be singled out by cooling rate testing, viscosity measurements, or a combination of these methods, intelligent treatment chemistries are possible.

Changes in Crystal Morphology

What changes in the crystal habit or morphology result from the addition of a crystal modifier?

One interesting characteristic of crystals is that they can form in a variety of shapes, which are due to the environmental conditions under which they form (see Fig. 10–6). They can be large or small, extend long distances or short, be well-defined or diffuse; in short, they can display an impressive array of forms. It is this variety of form upon which crystal modifiers are intended to take advantage.

Crystal Modifier Products

Throughout the previous discussions, the importance of population profiles has been emphasized. This emphasis is not an esoteric exercise; it has broad-reaching and practical application to the real world of wax problems in crude oils. There is a practical limit to the amount of solute (wax) that a solvent (liquid oil fractions) can contain and still remain liquid at a certain temperature. These limits are functions of the molecular weight of the wax and its concentration. Likewise, there are limits of the extent to which modifications by additives (e.g., crystal modifiers) can effectively alter the liquid-versus-solid characteristics exhibited by the waxy crude oil.

A chunk of iron is liquid at around 1,800° C, but below that temperature, it is solid. The same chunk of iron is placed in a 10% solution of hydrochloric acid, heated to 100° C for a couple of hours, and it dissolves. In the high temperature case, the iron (except for oxidation products) is still in the same chemical state, but in the hydrochloric solution it exists as Fe^{3+}. Thus, in order for the chunk of iron to exhibit liquid properties, it had to either be heated above its melting point or chemically altered. The solution of Fe^{3+} remains liquid at temperatures well below room temperature.

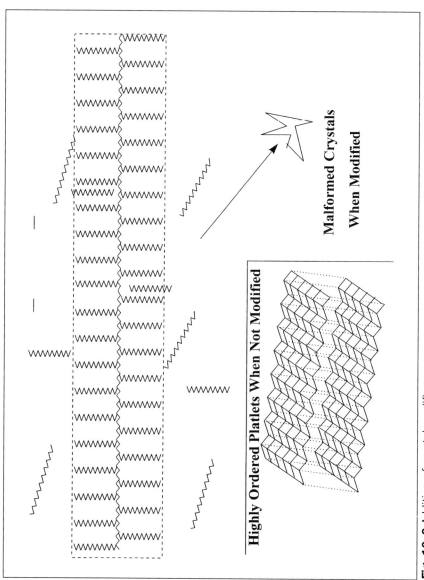

Fig. 10–6 Addition of crystal modifiers

A chunk of $C_{22}H_{46}$ normal paraffin wax is liquid at $44°$ C and solid below that temperature. The same chunk of $C_{22}H_{46}$ normal paraffin wax is dissolved in kerosene, and it is liquid at room temperature. The normal paraffin's chemical composition remains the same, but it is still liquid at room temperature.

The difference between these two systems is obvious: one changes chemically, while the other remains the same. Now, the same size chunk of normal paraffin wax is dissolved in 50% less kerosene; as the temperature is lowered, the mixture solidifies at $33°$ C. Thus, each time the solvent is reduced by 50%, the normal paraffin/kerosene mix asymptotically approaches the $44°$ C melt point temperature of the $C_{22}H_{46}$ normal paraffin.

Crystal Modifier Applications

The point of the paragraph above is intended to show that there are limits to which physical alterations of systems of solute and solvent can be extended. There are, however, modifications of the aggregate behavior of systems that can be achieved that are beneficial even at the limits of solubility. These benefits are primarily rheological, that is to say, crystal habits can be altered such that the bulk viscosity is lowered significantly by the addition of a crystal modifier.

Thus, the chunk of $C_{22}H_{46}$ normal paraffin can be treated with an effective crystal modifier to change its crystal structure, and, thereby, lower the sum of its aggregate interactive forces. Further, the solvated wax/kerosene mixtures can be similarly treated to affect even greater rheological improvement and very significant reductions of fluid viscosities. The effectiveness of the crystal modifier depends, to a major extent, on the type of aggregate-forming waxes that are present in the oil. Thus, the choice of a crystal modifier is key to the success of the treatment.

Crystal Modifier Synthetic Limitations

Crystal modifier chemistry is limited by several important factors. Among these are: availability of starting raw materials needed to approximate aggregation behavior of the type/types of problem waxes present in the oil, limited types of methods used for the modifiers synthesis, cost of synthetic methods and raw materials, polymer and copolymer chain conformation (orientation in space), and the reactivity and extent of polymerization.

Raw material availability is a major factor in the production of polymeric crystal modifiers. A substantial amount of monohydric fatty alcohol (single hydroxyl group) derived from the Alfol process is used for the production of crystal modifiers. The higher chain Alfol alcohols and olefins (C_{18+}) are produced as a by-product of this process. The Alfol process involves the polymerization of ethylene ($CH_2=CH_2$) using the Ziegler-Natta process, from which a Gaussian distribution of products ranging from C_4H_{10} through $C_{36}H_{74}$ is obtained. The major use of the Alfol products is for cosmetics, soaps, and creams, and the major fractions used are in the $C_{12}H_{26}$ through

$C_{16}H_{34}$ range. Thus the peak-range Gaussian distribution is adjusted to meet these product requirements.

The purity of these products is less than perfect because of the distillation processes used to isolate them, and the possibility for branching in the polymerization process. Only a few natural sources of these alcohols are available (C_{18+}), and although their purity and lack of branched structures are superior to the Alfol products, their cost is considerably higher.

Synthetic limitations are encountered when polymers and copolymers of these pendently substituted olefins and alcohols are pursued. In addition to the limited availability of the pendent substituents (alcohols and olefins), the type of reactions required to produce the pendency are also limited. These reactions typically include esterification of acrylates, methacrylates, and copolymerization with molecules like maleic anhydride. Thus, the functional group reactions are limited to easily accomplished reactions such as esterification. In the case where copolymers of ethylene and vinyl acetate are produced, high pressures and somewhat dangerous conditions are required to produce the desired product.

Some Wax Control Methods

Several methods have been employed for the control of wax problems, and some of these methods have strong theoretical basis for their use, while others are less well-rooted in theory. Crystal modification is well founded both in theory and application (see Fig. 10–7). The fact that paraffins crystallize in different geometric forms under conditions of cooling from identical solvents is a good indication that their morphology can be altered. And since paraffin morphology is susceptible to alteration, there is a theoretical reason to think that a crystal modifier will also affect a morphological change. The application of crystal modifiers to waxy crude oils has been successfully applied for several years to alter the morphology of the crystals formed.

Remedial Treatment Methods

Once waxes have deposited in pipelines, storage tanks, well tubing, and formations, another technique has been used to remove these deposits. Hot-oil treatment is one technique used by several producers, but this method suffers from at least one major drawback: a concentration of higher waxes left after treatment. This technique involves the circulation of heated production fluid through the transfer lines, well tubing, and formation to dissolve wax deposits. However, since the higher waxes tend to be less soluble at elevated temperatures, these treatments tend to concentrate the higher (harder) waxes in place.

Solvent washes are also used, where large amounts of solvents (e.g., xylene) are circulated through the system to remove deposits. These solvent treatments are more effective than the hot oil treatments and cause less damage, but are considerably more expensive. Additionally, both the hot oil and solvent-soak techniques are applied after the deposition of wax has occurred.

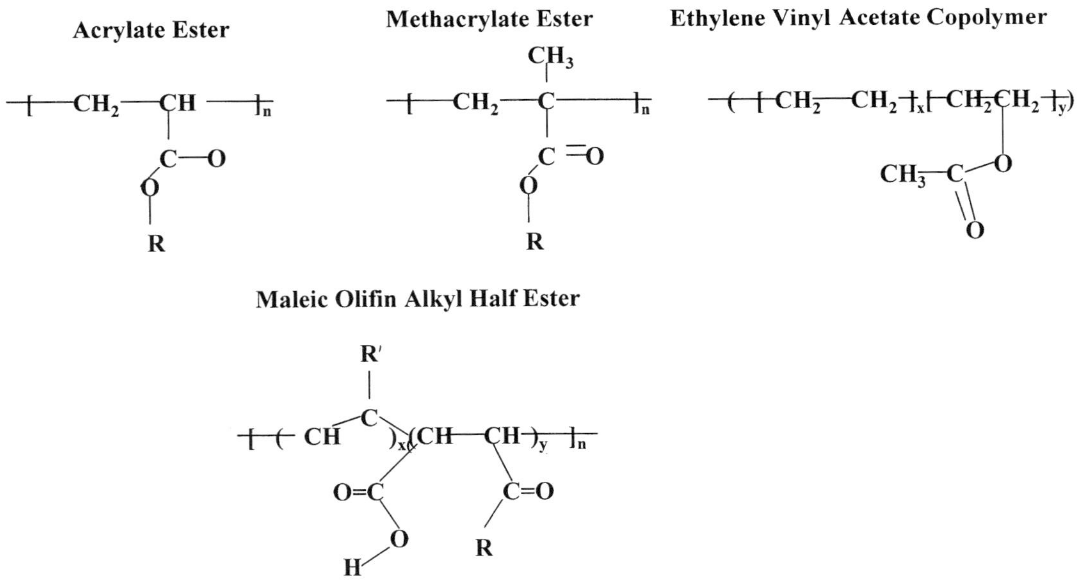

Where, $R = C_{16\text{-}50}$, $R' = C_{16\text{-}32}$, n =2-100+, x =4-15, y =4-15

Fig. 10–7 Some typical polymeric crystal modifier chemicals

Hot water and hot water with added surfactants have been used to remove deposits with a fair measure of success. But these treatments can only be used sparingly when applied to water-sensitive formations, and can also suffer from after deposit and formation sensitivity drawbacks. Although limited in application, hot water and hot water plus surfactant treatment can be used very effectively to remove existing deposits from less water-sensitive areas of production, storage, and transportation.

Mechanical Methods

Mechanical methods including scrappers, pigs, calipers, and high pressure nozzles have been used successfully once the production or transport facilities have had their functions interrupted. Again these methods are remedial in nature, and do not affect a change in the fluid properties of the crude oil. Attempts to employ microwave radiation to the tubing, and/or the transported production fluids in order to provide sufficient heat to melt the waxes has also be tried. This approach has had limited success and requires a substantial amount of polar dipole materials (since microwave radiation imparts rotational momentum) be present to affect reasonable temperature increases. Reports of the use of ultrasonic transducers have indicated very little positive effect under reasonable expenditures of energy.

Biotechnology

Microbial degradation has also been reported as having a beneficial effect on waxy crude oils. Inherent difficulties arising from the use of microbes in down-hole applications include:

- the need for heat-stable (viable) organisms
- reduced oxygen requirements by the microbe (anaerobic)
- heat-stable (noncoagulating) exo-enzymes
- organic-solvent-resistant cell walls
- the ability of the microbe to oxidize the terminal alkane to a carboxyl group before it can break down the paraffin by the acetyl-thio-coenzyme A reaction mechanism.

Several discussions about the mechanism of these microbial wax treatment methods have suggested that their effectiveness is most likely due to biosurfactancy. The discovery of thermophillic microbes, and their extremely stable enzyme complements, is encouraging and may offer an avenue of research. The existence of these microbes alone does not satisfy the problems mentioned above, but it does suggest the possibility for a genetically altered form that could be useful. The question of whether or not these altered thermophiles might be capable of selectively breaking down paraffin would also need to be addressed.

The ideal application of this biotechnology, once developed, would be the isolation of the specific agent (enzyme-protein) responsible for the degradation of the paraffin from the microbe grown under controlled conditions. Thus, the enzyme responsible for the breakdown is used instead of the whole microbe. Enzyme substrate studies have shown that the specific activity of the enzyme can be hundreds of thousands of times more active than the whole organism. Biotechnology is certainly an area for fruitful future research effort.

Supplemental Methods of Wax Control

Strong magnetic fields have been suggested and employed in certain cases, and some field personnel have reported success. The theoretical basis for such an approach suggests that forming waxes, when subjected to an intense magnetic field, will be displaced from their preferred crystal alignment (see Fig. 10–8). This disalignment then either prevents or interferes with the crystal's continued networking.

The method requires the installation of a strong magnet on a segment of the flow line. Usually the magnet is of the permanent type. The fluid passes through the segment, where it is subjected to an intense magnetic field. Several physical considerations must be taken into account if this method is to be applied successfully. Some of the considerations include: the temperature of the fluid in the segment, the differential temperature drop of the fluid across the segment, the fluid velocity through the segment, and the appropriate field strength. The temperature of the fluid must be known in order to affect those waxes undergoing aggregating at this temperature, and whether or not they are problem waxes. The need to know the differential temperature drop across the segment is crucial, since this drop is key to the transition state of the aggregating species.

Since an applied magnetic field is only in operation over the segment length, a knowledge of the fluid velocity is essential to the physical dimensions of the magnet. Thus, the exposure time of the fluid under appropriate conditions of temperature and pressure and the appropriate magnetic field strength necessary to alter these aggregation forces should be known. If all of these requirements are known and met, then this method could theoretically be applied successfully.

Summary

From the previous discussion of the treatment methods and limitations involved in producing chemistries capable of addressing the problems of wax deposits in the various production, refining, and transporting areas, it is clear some very stringent restrictions apply. In the area of prevention, the restrictions are considerably more confining. After-the-fact treatments are a great deal less restrictive, but they are also require forethought and planning. The stage of development of treatment methods is dictated by cost-performance criteria, and some of the more elaborate methods discussed are only practical when the payoff is high.

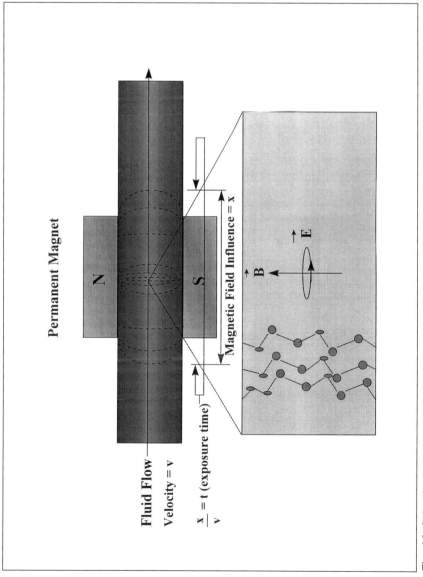

Fig. 10–8 Application of magnetic field for wax control

This chapter discusses the electromagnetic effects responded to and exhibited by aggregating and aggregated systems. It introduces the principle of superposition and how this principle can be used to derive a deeper understanding of the nature of complex systems. Piezoelectricity and superconductivity are discussed in relation to waxy deposits, and the phenomenon of electric monopoles is examined in relation to materials called electrets.

Additionally blackbody radiation is discussed in terms of its relationship to emissivity and a tie-in to time/space is suggested. The chapter was an attempt at making some of the more mathematically complex considerations of wax crystal formation a little easier to understand. Thus the concepts discussed in the sections on quantum mechanics and electrodynamics were applied to a description of the macroscopic physical manifestations of wax-forming systems.

Problems

10–1. Discuss the relationship of the imaginary number $(-1)^{1/2}$ in respect to time, temperature, and radiant energy.

10–2. The principle of superposition can be applied to the individual component contributions to emissivity. What advantage does this afford those interested in the physical behavior of aggregate systems?

10–3. If mono-poles arise from electric fields, why are magnetic mono-poles observed?

10–4. Give a rational explaination of whether or not piezoelectric phenomena would be expected from wax crystal aggregates.

10–5. What type of externally applied force instigates piezoelectric behavior?

10–6. What determines the vibrational frequency exhibited by piezoelectric crystals?

10–7. Can piezoelectric phenomena be related to superconductivity?

References

Barrow, Gordon M. *Physical Chemistry.* 2nd ed. New York: McGraw-Hill Book Co., 1966.
Handbook of Chemistry and Physics. 56th ed. Cleveland: CRC Press, 1975–1976.
Horowitz, P., Hall, Winfield. *The Art of Electronics.* 2nd ed. Cambridge: Cambridge University Press, 1994

Section III

Asphaltenes

11
Asphaltenes and Crude Oil

Asphaltenes Deposits

Asphaltenes are among the least understood deposits occurring in the oil field. They are thought to be the byproduct of complex hetero-atomic aromatic macro-cyclic structures polymerized through sulfide linkages (see Figure 11–1).

Porphyrins are common in nature and are part of a larger class of chemicals called cytochromes (colored bodies), which are iron-containing electron-transferring proteins. These proteins sequentially transfer electrons from flavoproteins to molecular oxygen, and they all contain iron-porphyrin prosthetic groups. The porphyrin ring is present in several biological systems and includes chlorophylls of green plant cells. Although many cytochromes have been highly purified, with one exception they are usually very tightly bound to the mitochondrial membrane and difficult to obtain in soluble and homogenous form. The exception is cytochrome c, which is very easily extracted from mitochondria by strong salt solutions.

According to Albert Lehninger,

The iron protoporphyrin group of cytochrome c is covalently linked to the protein via thioether bridges between the porhyrin ring and two cysteine residues in peptide chain, presumably formed by addition of the -SH group across the double bond of the 2- and 4-vinyl groups of the proto-porphyrin.

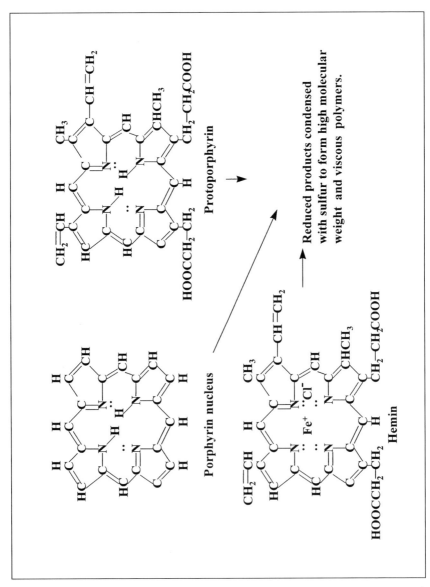

Fig. 11-1 Possible asphaltene origins

Thus condensation with other electrophiles (e.g., hydrogen sulfide) is possible and the product will consist of several repeating porphyrin units exhibiting very high molecular weights. The high degree of unsaturated ring carbons (double bonds between carbons) and the conjugated nature is responsible for extremely intense coloration of these species. An additional property of the porphyrin ring is its high degree of planarity derived from the fused unsaturated structure.

Coplaner Orbital Overlap or Pi Bonding

In the earlier sections of this book we talked about the roles of hydrogen bonding, London dispersion forces, and ionic interactions in aggregation phenomena related to complex hierarchical structures. An additional type of interaction is found to be operative for systems exhibiting high degrees of unsaturation, and this interaction is called pi bonding. The mechanism of this interaction involves the overlap of partially filled electronic orbitals between two conjugated (alternating double and single bonded atoms) molecular species.

The interchange of electrons between orbitals of separate molecules leads to a force of attraction between the two molecules. The electron interactions are quantum mechanical in nature, and a detailed description of the magnitude of these interactions requires quantum calculations be applied. However, as has been the practice so far, the intuitively understandable concept of blending orbitals will best suit our treatment of this phenomena.

Configurations of the molecules undergoing pi bonding must allow a sufficiently close approach for the electron clouds of the molecules to share electrons. The planer configuration of the porphyrin and certain other conjugated molecules allows these molecules to be stacked (see Fig. 11–2). If a comparison of relative strength is made between London dispersive attraction forces, hydrogen bonding, and pi bonding forces, the pi bonding strength would fall somewhere between them as in the following order: hydrogen bonding > pi bonding > inductive forces.

Polymeric Forms
Derived from Protoporphyrin

Given the information provided in the previous two paragraphs, and in particular the quote from Lehninger about the thioester formation of a polymer called cytochrome c, protoporphyrin provides a suitable structural analog to asphaltene. The fact that sulfur is involved in the polymerization is not altogether surprising, since sulfur radicals (single unpaired electrons) are very common in this element's chemistry. Sulfhydryl radicals are often formed from the heterolytic cleavage of hydrogen sulfide (a gas common to crude oils).

These sulfhydryl radicals are energetic enough to react with activated double bonds like those found on the protoporphyrin molecule. The double

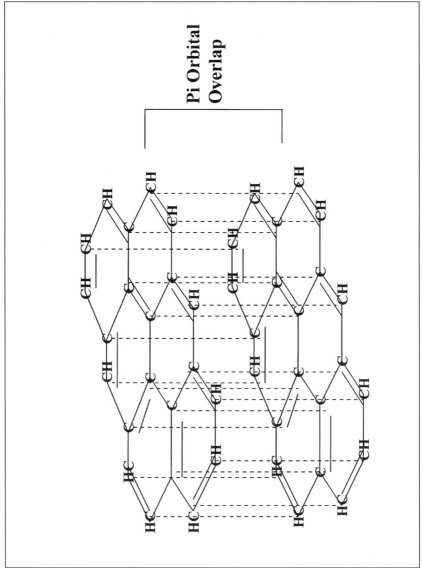

Fig. 11–2 Planer polyaromatic pi bonding

bond is activated by the electron-donating capabilities of the conjugated double bond of the macrocyclic protoporphyrin. The free-radical reaction is self-propagating, since the sulfhydryl radical decomposes to form a sulfur-terminated macroradical. The molecular weight is determined by several factors that include radical-radical termination, hydrogen abstraction from laibile molecules, and chain transfer reactions through abstraction of hydrogen to form a less reactive radical. A representative example of the perceived reaction and end product expected is illustrated in Figure 11–3.

The sulfide-linked protoporphyrin (pp-S-pp) is highly branched or cross-linked in the above example, but because of the possibility of ring resonance structures a secondary mechanism can be invoked to explain a significantly high degree of linearity. This mechanism involves the attack on the protoporphyrin by hydrogen sulfide at one of the vinyl carbons (ethylene substituent), a hydride shift from the sulfhydryl substituent, and radical formation of the cross-ring ethylene group. The resultant product of this reaction mechanism is a linear polysulfide-linked protoporphyrin of significantly high molecular weight. A proposed example of this reaction mechanism is illustrated in Figure 11–4.

The significance of the linear form of polysulfide-linked protoporphyrin is that it can be extended within solvation sheaths that result in significantly high crude oil viscosity.

Solvation Sheaths

Solvation sheaths for complex molecules such as the polysulfide-linked protoporphyrins, which will hereafter be called asphaltenes, can be viewed in much the same way as the bipolar stabilization of emulsions were described in Part I of this book. The asphaltenes represent a high degree of unsaturation and possess high aromatic character, while the surrounding crude is more aliphatic in character. Thus the solvation sheath presumably consists of molecules that exhibit both aromatic and aliphatic character. The use of the word "presumably" in the preceding sentence is intended to convey to the reader the fact that these structures have not been unequivocally identified. An illustration of the presumed solvated asphaltene is found in Figure 11–5.

It should be realized that the solvating species could consist of a large number of alkyl-substituted aromatic or hetero-aromatic molecules, as long as the possibility of pi bonding exists. Figure 11–5 represents a solvation sheath that consists of alkyl-substituted anthracene, phenanthrene, dibenzoanthracene, substituted benzopyrrole, and a host of different alkyl-substituted derivatives, which can be and often are present in the sheath. Thus a criterium for successful solvation is established, which requires the presence of both aromatic and aliphatic character in the solvating molecule.

Over the years, petroleum chemists and engineers have developed a nomenclature for the polysulfide-linked protoporphyrins and the group of solvating chemicals suspending them: they are asphaltenes, resins, and maltenes, respectively. These names are commonly used to refer to black, vis-

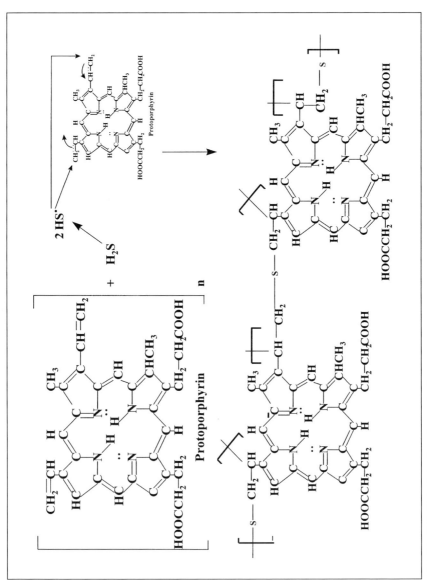

Fig. 11-3 Sulfide-linked protoporphyrins

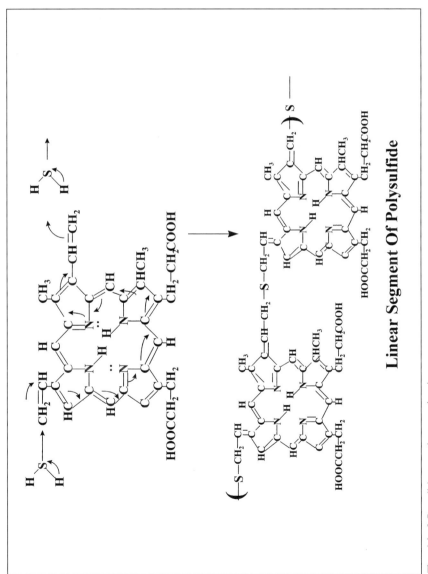

Linear Segment Of Polysulfide

Fig. 11–4 Possible reaction mechanism

217

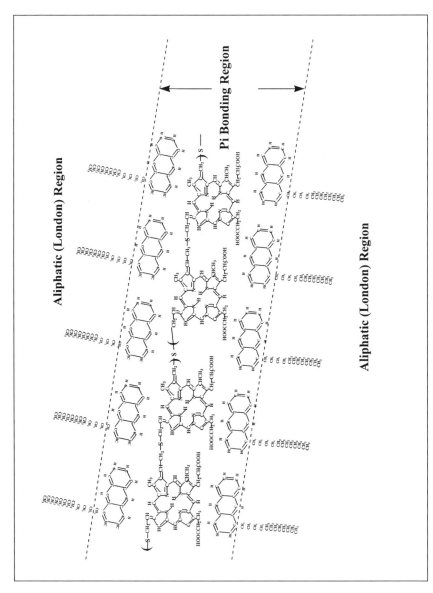

Fig. 11–5 Presumed solvated asphaltene

cous (low API gravity) petroleum crude oils. Later operational definitions will be used to describe some of the physical behavior exhibited by these materials.

Electrostatic Behavior of Asphaltenes

The high degree of electron delocalization associated with the asphaltenes, maltenes, and resins accounts for a relatively high charge dispersal effect that is distributed over the largely planer macrocyclic group. The overall sign of the dispersed charge is partially negative, and similar electronegative macrocycles are repelled; however, intermolecular blending of the bonding π orbitals does occur if the intermolecular distance is small and proximity is of sufficiently long duration.

This behavior is analogous to that of London forces of interaction between non-dipole containing molecules, with the exception that the resulting aggregate maintains its net electronegative character. The strength of the intermolecular association of these macrocyclic-conjugated systems is, as was mentioned, between that of the hydrogen bonding and London forces of induction. Thus these aggregates are shear sensitive, and once disturbed from their effective bond distances they accelerate rapidly from their solvation sheath.

Because they still possess an overall negativity, they can also be disrupted by electronic forces derived from magnetically induced frictional charges that arise from their motion with respect to an electron-rich sink (e.g., pipe wall or another asphaltene). These electrostatic interactions and shear sensitivity often result in stripping the solvent sheathing (resins and maltenes) from the asphaltene. As a result of the loss of the stabilizing maltenes and resins, the asphaltenes present a discontinuous surface to the surrounding crude oil and precipitate as insoluble, amorphous masses from the bulk fluid.

Asphaltene Destabilization

As mentioned earlier, asphaltenes are sensitive to shearing forces and electrostatic interactions. In this section we will try to gain an understanding of why these molecules are sensitive to electrostatic and shearing effects. Many naturally occurring and synthetic polymers are sensitive to shearing forces, and one of the main reasons they are involves the extended chain length of these polymers. Although the bonds produced by the reaction of vinyl monomers are covalent, the extended length of these polymers places significant stress on the internal bonds.

In symmetric molecules like linear polyethylene, the center of mass is equally divided on either side of the central carbon-carbon bond. The length of either of the two segments is equal, and when stress is applied at a point in the polymer chain, the momentum imparted to each of the segments is proportional to the mass and shear velocity acquired from the shearing force. If the magnitude of the shearing force is stronger than the carbon-carbon bond

strength, the bond will break and the polymer molecular weight will be reduced. Thus the longer the segments on either side of the applied shearing stress, the more likely the bond is to break.

Pi-bond orbital overlap and bonding is not nearly as strong as covalent bonding forces. Consequently bond shearing by relatively small forces can easily cause disruption. As London forces of attraction between fatty alkyl chains can be easily overcome by moderately small shearing forces, so too can pi bonding forces. The first pi bonding molecules to be removed are the resins and maltenes, and once these are stripped from the asphaltene, the core polymer becomes unstable because it is no longer surrounded by alkyl tails compatible with the bulk crude.

Unsheathed Asphaltene Cores

A secondary function of the resins and maltenes is the protection of the inner asphaltene core from external electrostatic influences and solvent incompatibility. The fatty alkyl substituents of the aromatic or poly aromatic resins and maltenes provide buoyancy and insulation for the asphaltene. When these stabilizing species are removed, the bare asphaltene presents an electronegative and solvent-incompatible surface to the bulk fluid, and begins to separate from the system as a solid. This generally occurs because the bare asphaltene core of one polymer chain can interact with that of another and form pi complexes of enormous size relative to the original parent. Other interactions are now possible with other surfaces that attract the electronegative core. Electropositive sites on pipe walls (or mineral deposits) can also act as collection points for the accumulation of these growing aggregates.

Metallocenes

Thus far asphaltenes have been described as polysulfide-linked protoporphyrin polymers, and little mention has been made of their powerful coordination abilities. Hemin and chlorophyll complexes are notable examples of the strength of these complexes, where iron and magnesium are bound to the central porphyrin nucleus of the protoporphyrin. These complexes are very common in biological systems and serve critical functions such as carrying oxygen to the blood in the case of hemoglobin, and electron transport for photosynthesis in plants. Hemoglobin contains an Fe^{2+} coordinated to the porphyrin nucleus, and functions as a ligand site for oxygen transport. The oxygen is reversibly coordinated with the Fe^{2+} and no oxidation or reduction of the iron occurs.

The anaerobic conditions under which biodegradation of these compounds occursare sufficiently severe to cause reduction to the protoporphyrin. Although complete reduction to the pure protoporphyrin is not expected, a significant percentage is thought to be reduced in the presence of less electronegative ions. Thus a more accurate picture of the peptized (resin and maltene sheathed) asphaltene is graphically depicted in Figure 11–6.

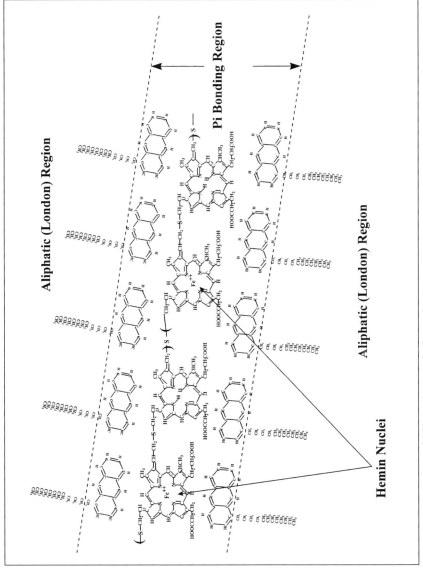

Fig. 11–6 sInclusion of FE²⁺ in the porphyrin nucleus

The inclusion of Fe^{2+} in the porphyrin nucleus is important, both in biological systems and in asphaltenes. In biological systems Fe2+ in hemoglobin is required for respiration, and hemoglobin and asphaltene structures are in a group of compounds known as metallocenes. The metallocenes involve bonding and antibonding p orbitals of aromatic nuclei with unfilled d-orbitals or hybrid orbitals of a central metal ion. This situation is depicted in Figure 11–7.

The arrangement of the hemin and protoporphyrin rings depicted in Figure 11–7 also applies to the polysulfide-linked asphaltene structures; however, it is questionable if the sheathed (peptized) asphaltenes conform to this bi-porphyrin metallocene structure. An interesting phenomenon associated with these metallocene structures is that many exhibit high magnetic susceptibility. The complexes $[Fe(CN)_6^{3+}$ and FeF_6^{3+} ...] in aqueous solutions yield a magnetic moment of 2.3 Bohr magnetons for the former and 6.0 for the later. This magnetic susceptibility is very much responsible for the interactions of asphaltenes with metal surfaces and other unsheathed asphaltenes.

Magnetic Susceptibility and Streaming Potential

If the structural composition of asphaltenes is comprised of metallocene complexes of Fe^{2+} or hemin nuclei copolymerized with protoporphyrins through sulfide linkages, a requirement of the Helmholtz theory of the double layer demands that charges be neutralized. Thus counterions must be present either in the porphyrin nuclei or at the interface between the stabilizing micelles (resins and maltenes). This precondition then suggests that fatty acid anions, amines containing unshared electron pairs, or oxides will be present at or near the interface between the asphaltene and the surrounding crude oil solvent.

Since moving charges and magnetic fields are known to induce magnetic and electric fields respectively, it is not surprising that the mobile hemin nuclei produce electronic effects in nearby susceptible surfaces. These induced fields not only affect the surfaces external to the asphaltene, but they also introduce a drag effect on the asphaltenes. The induced fields and drag-produced effects are described as streaming potential, and have formal mathematical descriptions that will not now be discussed. An interesting manifestation of streaming potential in low-conductivity systems is the production of static electrical discharges.

External Aggregate Asphaltene Destabilization

In the foregoing treatment of asphaltenes, maltenes and resins have been mentioned frequently, and a general aliphatic aromatic structure has been proposed to describe them. Although these structures are reasonable,

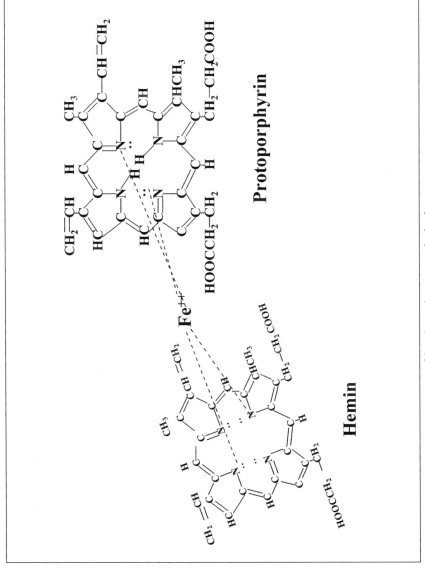

Fig. 11-7 Metallocene arrangement of hemin and protoporphyrin rings

they need refinement to further elucidate the mechanisms of their interactions with the asphaltene core. Much of the literature on asphaltenes suggests that the resins and maltenes (peptizing agents) are precursor molecules to the asphaltene. This possibility is quite likely, and reasonable candidates could be derived from the lactams of gamma amino acids or the pyrroles and pyrrolidines. Thus a fatty-alkyl-substituted pyrrole or fused pyrrole would serve very well as a bi-layer sheath. Figure 11–8 shows two possible candidates.

Although Figure 11–8 describes the fused pyrrole and proline as porphyrin precursors, they could also be breakdown products of a porphyrin parent. Thus the reports in the literature regarding the composition of the stabilizing or peptizing maltenes and resins would appear to be quite reasonable. The physical chemical properties of the pyrroles and fused pyrroles are quite interesting and exert strong interactive forces over the porphyrin nucleus. There is considerable transfer of negative charge to the carbon atoms in the pyrrole ring, which give it a high dipole moment $\approx$ 1.8 Debyes.

Because of the high nuclear p electron density, pyrrole is oxidized by air even more rapidly than aniline or phenol, leading to dark-colored resinous polymers. It is likely that this strong dipole acts to stabilize the hemin fraction of the asphaltene, while the aliphatic tail disperses it in the crude oil.

Figure 11–9 shows the polysulfide protoporphyrin surrounded by an alkyl substituted fused pyrrole, which is the structure we propose for the non-iron asphaltene. The inclusion of iron would favor this arrangement even more strongly because of the dipole moment present in the pyrrole. An important feature of the above-proposed arrangement is the positioning of the aliphatic tails of the sheath. This spinelike arrangement then allows the interaction of other aliphatic species such as paraffins, alkyl-substituted fatty acids, and alkyl tails of a wide variety of chemicals present in crude oil.

The interaction of these spinelike alkyl projections with similar alkyl groups results from London inductive forces. The more massive the interacting alkyl substituent, the more shear-sensitive the sheath becomes. When solvent molecules like propanes, butanes, pentanes, and hexanes are volatilized, the resulting agitation is often sufficient to disrupt the protective sheath (strip the maltenes and resins).

Summary

This chapter attempts to explain some of the chemistry involved with high aromatic, hetero-aromatic, and hetero-atomic macrocyles. It proposes plausible but hypothetical structures for components comprising a class of chemicals called asphaltenes, maltenes, and resins. An attempt was also made to explain the stabilization forces of the solvating intermediary resins and maltenes, and the nature of pi orbital intermolecular overlap. A scale of interactive strengths was developed to give an insight into the macro-aggregates association. Finally, a short discussion of the types of forces acting on the asphaltene aggregates suggested that shearing forces and electrostatic interactions could destabilize these aggregates.

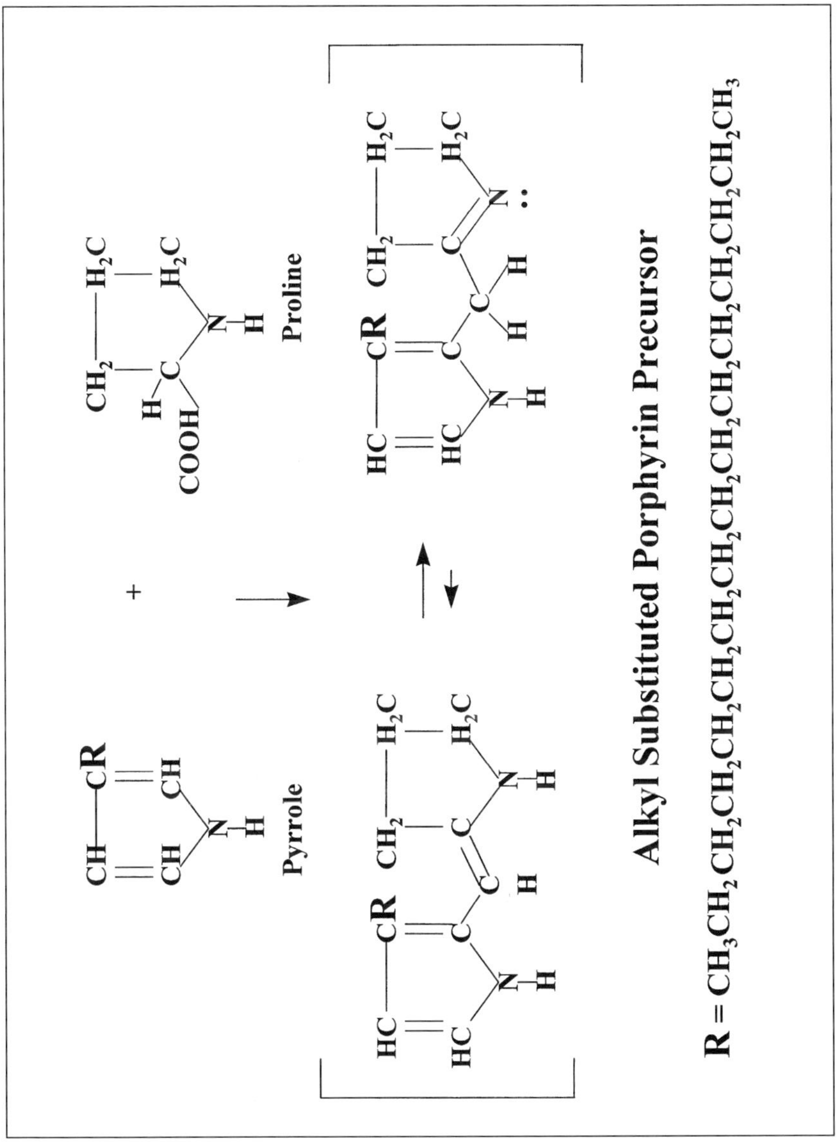

Fig. 11–8 Alkyl-substituted porphyrin precursor

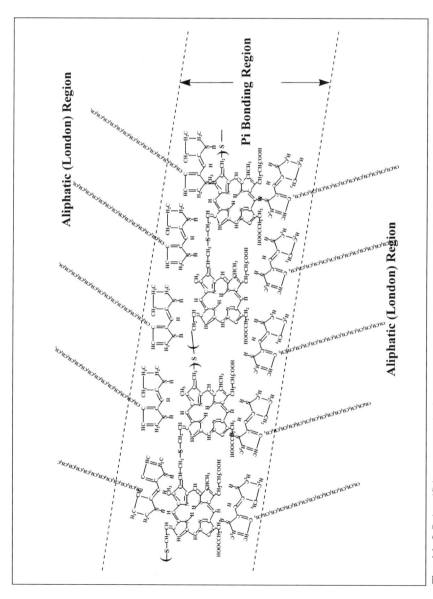

Fig. 11-9 Polysulfide protoporphyrin surrounded by an allkyl-substituted fused pyrrole

The preceding chapter also discussed the disruptive forces acting on the alkyl-pyrrole-sheathed polysulfide-linked protoporphyrin hemin structure (peptized asphaltene). Three main disruptive force types were implicated: shear stress, electrostatic, and London alkyl sensitization. The shear-stress forces imparted to the peptized asphaltene are small, but their very high aggregate weight make them susceptible to disruption by small shear forces. Streaming potentials exert drag effects on the core as the internal metallocene complexes induce charge in nearby surfaces, and the resultant core shifts act to strip the protective sheathing. Finally, the London attraction of external phase alkyl groups with the alkyl groups of the protective sheath increase sheath mass sufficiently to allow even minor agitation caused by the volatilization of light crude fractions to disrupt the sheath.

Problems

11–1. What type of interactions account for the strong protective forces found in asphaltene sheathing structures?

11–2. Rank the relative strengths of London forces, hydrogen bonding, ionic bonding, and pi bonding in order of increasing strength.

11–3. Explain how streaming potential affects the stability of the asphaltene aggregate.

11–4. Essay question: Propose an alternate structure for asphaltene aggregates, and justify the proposed structure on the basis of the behaviors exhibited by these aggregates. (Note, the structure described in this book is only a proposed structure.)

References

Barrow, Gordon M. *Physical Chemistry*. 2d. ed. New York: McGraw-Hill Book Co., 1966.

Handbook of Chemistry and Physics. 56th ed. Cleveland: CRC Press, 1975–1976.

Lehninger L. Albert. *Biochemistry: The Molecular Basis of Cell Structure and Function*. New York: Worth Publishers, Inc., 1970.

12

Bulk Behavior of Asphaltenes

Operational Definitions Versus Chemical Composition

It is instructive to note that the true structures of asphaltenes are not as unequivocal as the preceding chapter would lead one to believe. The oil industry has defined these materials on an operational basis, and this operational definition is more a description of physical behavior than actual chemical structure. Oil companies have known for years that asphaltenes are destabilized and precipitate from crude oil samples when added to multiple volumes of pentane. This observation has become entrenched in the literature and is accepted as the definition of asphaltene. Although this definition is unsatisfactory to chemists, it is valid in many respects to other professionals, such as petroleum engineers.

Thus screening processes aimed at determining asphaltene contents are conducted by a procedure that involves the dilution of the crude oil in multiple volumes of normal pentane. The job of the physical chemist is to rationalize this operational definition with chemical structures that would be expected to exhibit the observed behavior. Additionally, once a rational chemical structure is arrived at, it is the chemist's job to perform analytical

tests that further characterize the composition of the operationally derived sample (pentane insoluble fractions).

Because of the complex character of the macro-aggregates and the sensitivity to chemical alterations that are possible, these characterizations can be extremely complex. Thus, the actual structure or structures comprising asphaltenes will require the complete and unambiguous synthesis of these materials before complete characterization is accepted.

Indirect Evidence for Asphaltene Composition

Many times the behavior of chemical systems when acted upon by external influences can be invoked as deductive mechanisms for understanding the chemistry of components within the system. Such deductions are often retrievable from the processing of crude oils after their production. One such process is the extraction of heavy vacuum tower bottoms in the refinery. This process involves the N-methyl pyrrolidone extraction of these heavy bottoms products to remove the polar species from the nonpolar in preparation for further refining processes.

The N-methyl pyrrolidone combines with the black tar bottoms and allows phase separation of the nonpolar paraffins. The paraffin bottoms are sent to the dewaxing process, and the pyrolidone fractions are sent to the asphalt distillation unit for retrieval of solvent and preparation of asphalt. Given this information, what type of chemical structures would be expected to partition so readily into the polar N-methyl pyrrolidone phase? Or more importantly, would the proposed structures developed in chapter 11 be expected to yield to this solvent treatment?

In order to answer these questions, it is important to understand some of the properties of the solvent. N-methyl pyrrolidone is a high-dielectric aprotic solvent, which solvates by dipole-dipole interaction with the solute. Aprotic solvents are held less strongly than protic solvent molecules, and since they are held less strongly, the solvation activation energy is less.

An example of this is the rate of displacement of an iodide ion by a chloride ion in methanol (protic solvent) versus dimethylformamide (DMF, aprotic solvent) where the rate in the DMF is more than 1,000,000 times faster than in the methanol. For a given aprotic solvent, the attraction between solvent molecules and nucleophile increases with the polarizability of the nucleophile, and hence with its size. Thus the proposed structures are compatible with the solute solvent behavior described above. Figure 12–1 shows some of the solvent features of N-methyl pyrrolidone.

Field Problems with Asphaltenes

Asphaltenes are responsible for many of the field problems experienced by production companies. The discussion on the proposed structure of asphaltenes and the associated peptizing agents has explained much about

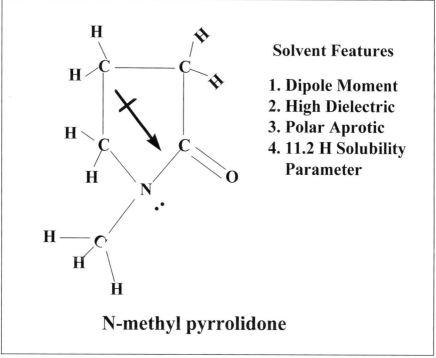

Fig. 12–1 N-methyl pyrrolidone

the physical chemical behavior of these systems, but little information has been given about their behavior in actual field systems. The deposition of asphaltenes has been observed in a large number of locations in producing wells and associated transfer and storage facilities. Some of these depositions occur at locations of particular interest, which include well chokes, casing perforations, formation rock, the vacuum side of pumps, storage tanks, and well tubing. With the exception of storage tanks, each of these locations has something in common, and that something is shear and drag.

In the discussion on the sensitivity of the asphaltene aggregates in chapter 11, the stripping of the peptizing agents from the core asphaltene was proposed as a reason for precipitation. But what about the storage tank? Again, the discussion of London alkyl aggregation and consequent sheath destabilization helps to answer the question. The quiescent conditions prevailing in storage tanks favor accumulation and aggregation of alkyl fractions with the alkyl exterior of the peptized asphaltene. Over a period of time the aggregations have become so massive that the mild agitation accompanying light-end loss becomes sufficiently strong to desheath the peptizing shell.

In addition to deposition, other problems are the result of asphaltene content in crude oils. A significant problem is that of viscosity, and the associated difficulties of production rate, pumping energies, and transfer rates. High asphaltene crude oils are generally high in viscosity and difficult to handle. The molecular weight of the asphaltene and its high dispersion throughout the crude mix determine the viscosity of the crude oil.

Although we have just begun to elucidate the structural composition of the asphaltenes, we have not yet determined the extent to which it is polymerized. Thus, viscosity effects associated with this molecular weight remain somewhat of a mystery. However, some intelligent guesses are possible. At least two things must be known in order to determine the effect of asphaltene molecular weight on crude oil viscosity: asphaltene concentration and average chain length of the polysulfide-linked constituents. We can determine the quantity of asphaltene by the pentane insoluble method, but determining the polysulfide constituents is more difficult.

Assume that the protoporphyrins and hemin constituents arise from decomposed plant and animal systems. Assuming also that the proportion of these substituents is small compared to paraffins, fatty acids, and other biomolecules, then it is a safe bet that these molecules will have a molecular weight that is inversely proportional to that ratio. And as a result of this hypothetical experiment, molecular weights of approximately 500,000 to 1,000,000 grams per mole are expected (hemin and chlorophyll ratios in micrograms per gram). These molecular weights would most certainly pose viscosity problems even at low concentrations.

Acid Effects on Asphaltene

Field personnel often conduct acid jobs on wells that have scaled up with inorganic salts. These acid jobs are usually conducted by back flushing the wells with hydrochloric acid solutions. Very often the acid treatment of an asphaltene-containing well will result in coagulation of the asphaltenes and an even more serious production impediment will result.

Over the years the well-service companies that conduct these acid cleanups have developed chemical additives that are intended to ameliorate the problem of sludge formation. A great deal of proprietary chemistry is involved, and a listing of specific chemicals would be out of the scope of this book. The success these companies have had with additives for asphaltene sludge control has been less than spectacular. The types of chemistries are extremely varied, and the concepts of how the additives act to prevent sludge are not fully rationalized. However, many of the additives are complexing agents with multiple ligand sites such as EDTA, citric acid, tartaric acid, biuret, and glycols.

The addition of these complexing agents does appear to have a positive effect on the reduction of sludge when added to the acidizing package. This effect is attributed to the sequestration of acid soluble iron salts that are found in high concentration in the spent acid wash. However, given the proposed structure for asphaltene described previously, it is not surprising that sequestration of iron salts would reduce sludge arising from the acidization process.

One additional factor is operative in the destabilization of asphaltenes by strong acid solutions, and this factor arises from the fact that pyrroles are slightly acidic with a $pK_a = 15$ compared to phenol, $pK_a = 9.9$. This is one factor for the destablization of the peptizing sheath. (An interesting side note about the behavior of protoporphyrins is that when they are reacted with alkaline solutions of ferric chloride, hemin is formed.)

Artificial Causes of Asphaltene Deposition

As mentioned earlier, certain techniques intended to enhance the recovery of petroleum result in exacerbation of the problem of asphaltene deposition. A particularly striking example of this is true is in the widespread usage of CO_2 flooding. This technique results in significant amounts of asphaltene deposition when the formation being treated contains heavy crude oils. Although less common, the use of nitrogen for enhanced production also produces asphaltene deposits. The nitrogen flood method, however, produces less sludge than the carbon dioxide injection method.

What factor in these two methods could account for the quantitative differences in the amount of sludge formed? The conditions under which injection of the flood gas takes place and the conditions the gas encounters during its torturous journey to the fluid matrix determine the condition and composition of the gas when it arrives. Nitrogen is quite stable at high pressure and significantly high temperatures, whereas carbon dioxide can and does undergo chemical alteration under similar conditions. Thus, the percentage of carbon monoxide (CO) reaching the reservoir can be significant.

As mentioned above, the hemin function in respiration is transportation of oxygen to the tissues. It is also commonly known that carbon monoxide is more strongly bound to hemoglobin than oxygen, and that carbon monoxide is a strong complex with hemoglobin, of which hemin is a major component. This carbon monoxide thus produced can complex with the hemin in the asphaltene core. This complex probably does not destabilize the polysulfide-linked hemin protoporphyrin core. It is more likely that the by-products of carbon dioxide break down to carbon monoxide and carbonic acid that cause the destabilization. This is because the peptizing agents (alkyl-substituted pyrroles) are acid sensitive, and water-coordinated carbonic acid affects the destabilization.

Oil Field Treatment Methods

Asphaltene depositions are common in a very large number of producing wells throughout the world, and their cost to the producers is significant. Most crude oil production companies use xylene washes for the removal of asphaltene deposits from wells, transfer lines, and storage facilities. This remedial procedure requires considerable investments in time, solvent, and lost production. Often the depositions are so severe that the well's production rate falls from thousands of barrels per day to zero.

The rate of decrease in production depends on the rate of deposition, but in severe cases production rates can drop to only a fraction of the unobstructed rate within days. Thus methods to prevent these depositions are an important part of a production company's strategy. Many service companies have been actively involved in the research and development of products to address the problem of asphaltene deposition, but these efforts have been hampered by the extreme complexity of the problem. Much of the difficulty encountered by these research efforts involves the lack of a clear understanding of the chemical nature of the deposits, and the specific chemistries needed to effectively treat the problem. Therefore, production companies have had to rely on solvent wash procedures and the ingenuity of production engineers to address these problems.

Over the years, many attempts have been made to actively interfere with the deposition of asphaltenes, and only a very limited progress has been made. Three of the very limited number of chemicals that have been introduced into the problem crudes with partial success include aromatic fatty sulfates, fatty amine aromatic fatty sulfate salts, and aromatic amines. The degree to which this class of chemistry is successful in preventing deposits can be described as extremely limited at best, and even harmful at worst. Therefore, continuous treatment programs have not been applied in many cases. However, the addition of these chemicals to solvent wash systems has enhanced removal effectiveness.

Production Factors Affecting Asphaltene Deposition

The production engineering staffs of many oil companies have been responsible for the major progress in the prevention of asphaltene deposits. The success has been due to production procedures, system equipment design, and remediation procedures implemented by field engineers. Production procedures are extremely important for optimization of production from crude oil reservoirs. Production rates, wellhead pressure, fluid and gas turbulence control, perforation depths, fracturing control, tubing design, and transfer-line design are all areas in which engineers have succeeded in controlling the deposition rates of asphaltene deposits.

Although physical designs have been the most successful methods for the control of asphaltene deposition, there is a practical limit to the extent to which design modifications can be implemented. For example, producing a well below the bubble point pressure (e.g., the pressure at which the most volatile crude fractions begin to escape from the liquid) is known to reduce asphaltene deposition, but in some cases this pressure is below what is required to lift the crude.

Obviously this situation is unacceptable from any point of view, so production must occur above the bubble point. An interesting side note to this situation involves the natural tendency of asphaltenes to form constrictions within production conduits that act as chokes to maintain flow rates, which are below the bubble point of the fluids. This effect is driven by the asphaltenes' natural tendency to seek an equilibrium point of effective dispersion in the crude oil matrix.

Asphaltene Deposition Control by Treatment of Other Problems

In the preceding chapter about the proposed composition of asphaltenes, maltenes, and resins (and their interaction with surfaces and surrounding chemicals) fatty alkyl interactions were seen to play a role in the stability of the asphaltene aggregates. These interactions include the association of long chain paraffin waxes with the aliphatic tails of the protective sheath of the asphaltenes.

The association of the paraffin waxes with the asphaltenes has long been known to reduce the tendency of crude oils to exhibit wax networking phenomena. Thus crude oils with high percentages of wax accompanied by asphaltenes exhibit lower pour points, but higher viscosities due to the high molecular weight of the asphaltenes. Additionally when asphaltene deposits are examined, they are found to be in close association with wax. Very often the percentages of each component, wax and asphaltene, are found to be nearly equally divided. Thus confusion arises as to which component actually causes the deposition problem, and treatments directed at the misunderstood problem sometime result in favorable results.

The introduction of wax crystal modifiers often inadvertently affects a solution to an asphaltene problem by reducing the instability introduced in the peptizing sheath of the asphaltene by wax interaction with the externally directed alkyl tails of the sheath. Conversely, the introduction of one of the limited number of effective asphaltene dispersants also inadvertently effects a solution to some wax problems.

Fluid Transport Equipment

The high sensitivity to shear destablization exhibited by asphaltene aggregates requires that turbulence be minimized in the production and transport of crudes containing asphaltenes. Thus the use of low-shear pumps is advisable in transporting asphaltene containing crude oils. Sudden expansions in line construction and large numbers of valves and other fluid-diverting methods should be avoided, since they add turbulence to the fluid. Most of these recommendations are commonly practiced by oil producers and have evolved to a high level of efficiency over the years by methods of trial and error.

One factor that affects the deposition of asphaltenes may not seem important on the surface, but it is more important than commonly recognized: corrosion. Given the proposed structure for asphaltene aggregates, it should be recognized that the presence of oxidized metals can pose a serious threat to the stability of the asphaltene aggregates. The transition metal complexes of protoporphyrin can form with great facility, and the stability of the aggregates is largely determined by the nature and extent of the resulting complex. Thus, corrosion control can also be useful in the prevention of asphaltene depositions. Since it is not practical to construct well tubing, transfer lines, fittings, or storage facilities from corrosion-resistant metals, chemical additives that act to prevent corrosion of mild steel can be helpful in the prevention of asphaltene problems.

Summary

The preceding chapter discussed the operational definition of asphaltene versus the proposed chemical structure. The operational definition is based on the dispersibility of the macro-aggregates in a pure hydrocarbon solvent (typically pentane). The chemical composition was derived from direct and indirect methods, and seeks to explain physical behavior that is consistent with the operational definition. The process of deduction is heavily employed and involves the extrapolation of known physical behavior in processes such as solvent washing of vacuum tower bottoms with polar aprotic solvents.

Further deductions are made from the behavior of asphaltene-containing oils when subjected to mineral acid treatments of producing wells. Inferences were also drawn from the effectiveness of sequestering agents used for sludge prevention during the acid treatment of wells. Finally, a rationalization of the behavior of asphaltenes in the presence of artificial drive stimulants was presented, which appears to be consistent with the proposed structure. Thus the entire chapter is devoted to the justification of the proposed structure for asphaltenes.

In the preceding chapter we also discussed the field treatment methods for the prevention and removal of asphaltene deposits occurring in oil wells and surface equipment. It was noted that the majority of successful techniques for asphaltene deposition control are the production techniques developed by field engineers. It should also have become evident to the reader that the chemical methods of deposit prevention are often the result of misapplication of treatment chemistries. Additionally the methods for the removal of deposited asphaltenes is still approached through the use of solvents and chemical dispersants. Thus the area of asphaltene deposit prevention should be considered an area of continued emphasis and a fruitful area for specialty chemical company research and development.

References

Acheson, R.M. *An Introduction to the Chemistry of Heterocyclic Compounds.* 3d. ed. New York: John Wiley & Sons, 1976.

Barrow, Gordon M. *Physical Chemistry.* 2d. ed. New York: McGraw-Hill Book Co., 1966. Rodriguez, Ferdinand. *Principles of Polymer Systems.* New York: McGraw-Hill Book Company, 1970.

Seymour, Raymond B. *Introduction to Polymer Chemistry.* New York: McGraw-Hill Book Company, 1971.

13
Asphaltene Testing Methods

Solvent Testing

Testing for asphaltenes is a problem in many respects, since asphaltenes are operationally defined by solubility in multiple volumes of normal pentane. Because of this poorly characterized and inadequate physical chemical definition, test methodologies can vary considerably from laboratory to laboratory.

Several questions arise when solvent tests are performed on crude oil samples. Should the crude sample be filtered to remove extraneous solids before being introduced into the pentane? If the sample is prefiltered, will some of the asphaltenes be removed? Should an effort be made to remove the wax fraction prior to solvent testing? Should the filtration be conducted on heated samples? Should the sample first be solvent tested, the precipitate filtered, and ashed in a muffle furnace to determine inorganics? Will prefiltration destabilize the asphaltene aggregates by shear effects? And so on.

The end result of all these questions is more questions, and the answers are often as subjective as the operational definition. Thus standardization of all testing is required; however, the standard methods are just as illusionary as the operational definition. Herein lies a problem for those seeking a fundamentally supportable testing procedure. It also leads to the confusion surrounding the structural elucidation of the offending materials, and complicates the design of chemistries that could have application in deposit reduction. Considering this state of confusion, it is quite reasonable to make assumptions based on the indirect evidence of asphaltene structure and extend these assumptions to testing procedures.

Core Testing Procedures

Core test methods can be very effective in determining the deposition tendencies of organic materials, and if they are conducted with the appropriate precautions, they can help answer some questions about the behavior of asphaltenes (see Fig. 13–1).

The extent to which this test methodology is employed is not known, but it does not appear to be widespread. The concept is simple, however, and the testing results are looked upon favorably by oil company personnel. Whenever possible core samples are obtained from the producing formation, and all testing is performed on these samples. In addition to the core sample, samples of the asphaltene deposits from the same formation are tested on the core.

Since xylene is the solvent most often employed for well cleanup of asphaltene deposits, the sample is suspended in xylene and pumped through the core. The pressure required to force the dispersed asphaltene xylene mix can be quite high (100 psi or higher) so it is necessary to use an HPLC pump (high pressure liquid chromatography pump). After several core volumes have been displaced, the xylene mixture is replaced by pentane, and circulated while monitoring pressure and absorbance. The addition of pentane is intended to cause precipitation of asphaltenes in the core and thereby cause a pressure increase. The spectrometer is standardized on the initial xylene mixture and compared to a Beer's law plot of concentration versus absorbance. Once the core is fouled with asphaltene, xylene or xylene plus chemical is added to test the removal efficiency of the chemistry. Other testing can be performed, such as adding a chemical to the original xylene asphaltene mix, or to the pentane to determine the effectiveness of deposition inhibitors.

Thin Layer Photometry

One technique that has been reported involves the use of a high-pressure dual window quartz cell through which the live oil (sample retrieved under reservoir conditions) is circulated. The pressure is gradually reduced from reservoir conditions while a light source passes through the circulating sample to a photodetector. When precipitation begins, the light is scattered by the particles of forming organic solids, and the degree of scatter is a measure of the precipitate. This method allows a combination of visual observation, photometry, and video recording during the test. The use of live oil is a very important part of this test methodology. By using live oil, the test procedure avoids the problem of a previous deposition history and more accurately represents the behavior of fluids as environmental conditions change.

There are also some drawbacks to this testing procedure, not the least of which is the provision of live oil. Since live oil must be retrieved under conditions existing in the reservoir, complicated procedures for collecting the fluid are involved. The producing oil company must stop production, remove tubing obstructions, and lower a pressure container into the well to collect

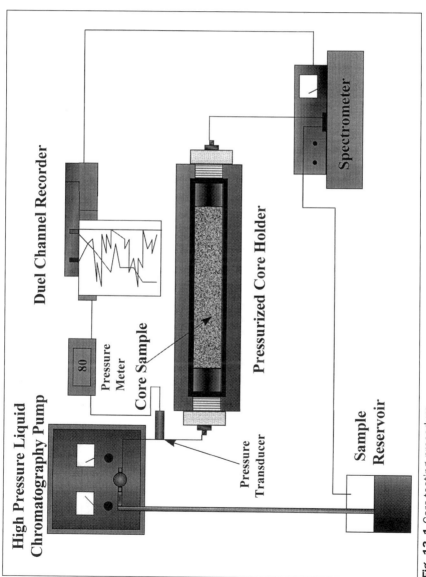

Fig. 13–1 Core-testing procedure

the sample. These collection procedures are expensive and difficult, so the producers are reluctant to provide samples. However when this test method is used, the results appear to be very reliable as a means of determining deposit tendencies.

Figure 13–2 shows the arrangement of the asphaltene deposition device. It is also interesting to note that this device is the same as the cloud-point device shown in the section on waxes. The difference in testing methodology between cloud-point tests and asphaltene deposition determines the results obtained. The cloud-point test observes the behavior of the circulating fluids under conditions of decreasing temperature and constant pressure, while the asphaltene test observes the behavior of fluids under decreasing pressure at constant temperature.

Size Exclusion Gel Permeation Chromatography

The idea of separations based on size seems to be a reasonable approach to the characterization of asphaltene aggregates, since the molecular weight of the asphaltene core can be very high. The concept of size exclusion is based on the idea that higher molecular weight materials will be excluded from high surface area resin beads that have well-defined channels of known size that accommodate small molecules. These resins can be obtained with a variety of pore channels and placed in a column in layers of increasing pore size.

The sample is placed on the column, and a mixture of solvents are then added to drive the sample through the column. As the sample proceeds through the column, the smaller molecules are incorporated into the pore spaces of the layered resins, while the higher weight materials continue to progress through the column. This procedure is usually assisted by the use of a high pressure pump that provides for a reasonable flow rate. Although the pressures used tend to be high, the flow rates are low; consequently, the shear forces also tend to be low. Thus, the appropriate choice of solvent/solvents used as eluents is extremely important for the success of this method.

This technique has many advantages, one of which is the isolation of reasonably large samples that can be used for further characterization. One disadvantage of this method is that it will almost certainly remove the sheath from the asphaltene aggregate, making it difficult characterize these materials.

Proposed Microwave Tests

Microwaves radiate energies that occur at levels that are of the same magnitude as the rotational energies exhibited by molecules possessing a dipole moment. Given the proposed structure for asphaltene aggregates from previous sections, absorption of microwaves would be expected. The unfortunate fact that large molecules possess so many internuclear distances that

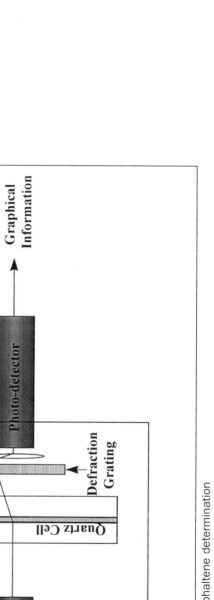

Fig. 13–2 Light-beam asphaltene determination

the three-dimensional moments of inertia cannot be sorted from them makes indirect measurement necessary. Additionally the interactions of molecules in liquid or solid phases confuse spectral interpretations, and microwave studies are generally conducted on gas phase samples.

Although the foregoing limitations would appear to prevent any application of microwave analysis to samples of asphaltenes, some interesting behavior of these aggregates make microwave studies possible. Asphaltene can be atomized (carried by sufficiently energetic gasses into head spaces). This behavior is observed in gas processing plants, where asphaltene deposits occur above the liquid level in compressor overheads. Because of this phenomena, the intentional atomization of asphaltene aggregates can be subjected to microwave radiation studies. If the gas used for atomization possesses no dipole moment, the microwave spectra of the atomized asphaltene will be the absorbing species.

The problem of high-molecular-weight material possessing a multiplicity of internuclear distances still needs to be addressed. By adding gas phase molecules possessing varying dipole moments to the atomization stream, indirect spectral measurements could be made. Thus the atomized asphaltene aggregate absorbing microwaves in a particular frequency range would be selectively masked by the addition of known wavelength absorbing polar molecules.

Proposed Ligand Replacement Method

Selective extraction of the Fe^{2+} by the addition of phthalonitrile is possible, since the complex formed under temperatures of approximately 200° C gives a phthalocyanine structure. This phthalocyanine structure is analogous to the hemin structure but contains four additional nitrogen molecules replacing the methylidyne groups. Figure 13–3 shows the structure of the phthalocyanine and the perceived interaction with the hemin fraction of the asphaltene aggregate.

After the selective extraction of Fe^{2+} is accomplished, the sample can be washed with acetonitrile to remove the soluble phthalonitrile, and titrated to a potentiometric endpoint using standard copper sulfate solution.

Designing Asphaltene Deposit Control Chemicals

Much of the chemistry developed to address surface phenomena has, in many cases, evolved as a process of trial and error rather than as the result of profound insight into the interactions producing the surfaces. However once effective treatment chemistries have been developed, rational arguments based on fundamental concepts can be used to explain the interactions. In fact, what incremental progress is made is generally due to a more thorough understanding of the chemical and physical processes taking place

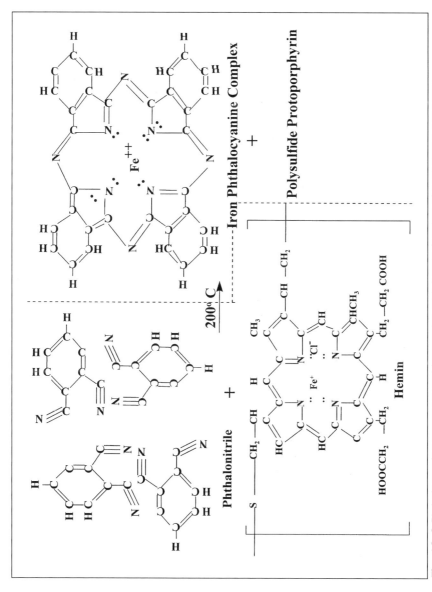

Fig. 13–3 Phthalocyanine structure and interaction with hemin fraction of asphaltene aggregate

in the systems being studied. Thus, approaching a physical chemical problem with only rudimentary understanding of it can be a very large impediment to arriving at adequate chemical treatment methods.

This has been the case with respect to the behavior of asphaltenes. Often approaching the solution to a problem requires that the chemist ignore the commonly accepted descriptions of the system and work from a framework built from his or her own experience with the problem. Because the asphaltene literature is confusing, and contradicts many of the experiences of this chemist, it was necessary to rebuild a model that was consistent with my own experiences. As a result the proposed structures developed in chapters 11 and 12 are the models that will be used to design some possible synthetic products to address the problem of preventing asphaltene deposition. This will begin by the examination of the structures and elucidating reactive sites that can be used.

Chemical Handles

The design of chemistries capable of affecting the behavior of macro-aggregate structures is often approached by drawing from analogous behavior in other systems. An example of this kind of inferential reasoning is illustrated by the use of quasi oil-soluble/ water-soluble materials in the treatment of emulsions. Emulsions are composed of bipolar molecules that exhibit this biphase solubility, and because they do, a reasonable approach to their resolution consists of the use of materials that exhibit the same qualities. Thus the property of bi-phase solubility can be considered a chemical handle that provides a means by which the emulsion may be resolved.

Approaching the development of chemicals that possess properties similar to those of the proposed asphaltene aggregate involves the same inferential reasoning. What chemicals will interact with the sheath molecules or the core molecules to impart additional stability to the aggregate? What chemicals could alter the core and enhance its dispersion? What type of chemical could cause the asphaltene aggregate to be coiled instead of extended in the crude oil? Can typical parts per million dosage levels of chemicals accomplish stabilization, or coiling of the asphaltene core, to produce a lower system viscosity?

Examination of the proposed asphaltene structure reveals that alkyl-substituted pyrroles surround the asphaltene core. If these pyrroles are displaced by a polymeric analog, will the polymeric analog hold the asphaltene core more strongly than the individual alkyl substituted pyrroles? Further, since the alkylated pyrroles are weakly acidic, will a somewhat stronger acid analog bind more completely with the asphaltene core? At this point, it should be obvious that the process of asking the last two questions has provided a jumping off point for a synthetic approach. Figure 13–4 shows a candidate that came into focus by the posing of the two questions above.

The possibility of the polymer formed in Figure 13–4 being arranged as a protective sheath around the asphaltene core is determined by the confirmation of the polymer backbone of the poly vinyl octadecyl pyrrolidone. If the

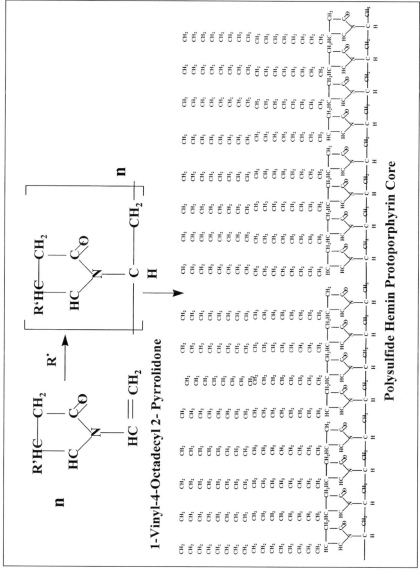

Fig. 13–4 Possible polymeric asphaltene stabilizer

intended protective polymer can take a spiral form, and the unshared pair of electrons of the nitrogen can interact with the core, then a protective sheath could be formed. This protective sheath would be considerably less sensitive to stripping forces, because of its increased molecular weight and large number of delocalized electrons on the nitrogen of the pyrrolidone ring.

Chemical Treatment Versus Reaction

A very important aspect of chemical treatment involves the alteration of the physical behavior of a system by the addition of very small quantities of chemical under conditions of temperature and pressure that prevail in the field. Chemical reactions, on the other hand, require specific conditions that are seldom available in actual field situations. Therefore, any attempt to alter the physical state of a system is constrained by the prevailing conditions. In addition to these constraints, the complex components present in crude oil must also be considered, and the product must zero in on the problem. Thus the chemistry developed must have the power to affect a change in the physical system by interacting with the offending material at low levels, and under less-than-ideal conditions.

In chapter 12 we saw that complexes of phthalocyanine could be used to selectively extract the Fe^{2+} from the hemin fraction of the asphaltene core. Because of the strength of this complex, it seems that the precursor molecule phthalonitrile could be useful in addressing some of the physical problems associated with the asphaltene aggregates. The use of complex formation for altering the asphaltene aggregate was suggested in chapter 12 in combination with well acidizing programs. Thus the processes of development benefits from a knowledge of practices and procedures that have been effective in seemingly unrelated areas.

The question of whether or not the extraction of the iron from the hemin group of the asphaltene core will act to stabilize the aggregate is still not answered, but a plausible argument for potential stabilization has emerged. By removing the Fe^{2+} from the asphaltene core, streaming potential forces associated with the iron complex or hemin have been reduced. The reduction of streaming potentials resulting from the inclusion of iron in the core should enhance the stability of the aggregate. Figure 13–5 indicates why removal of iron from the hemin could possibly lead to lower streaming potential forces.

Figure 13–5 illustrates the possible extraction of iron from the hemin group of the asphaltene core by the formation of the phthalocyanine complex formation from the addition of phthalonitrile. The resultant polysulfide protoporphyrin core loses its surface charge, and the phthalocyanine iron complex is readily dispersed in the crude oil matrix.

The polysulfide protoporphyrin core still presents a problem because of its high molecular weight and matrix extensibility, which is exhibited as high viscosity. Thus, some method of cleaving the polysulfide links under mild production conditions would appear to be an approach.

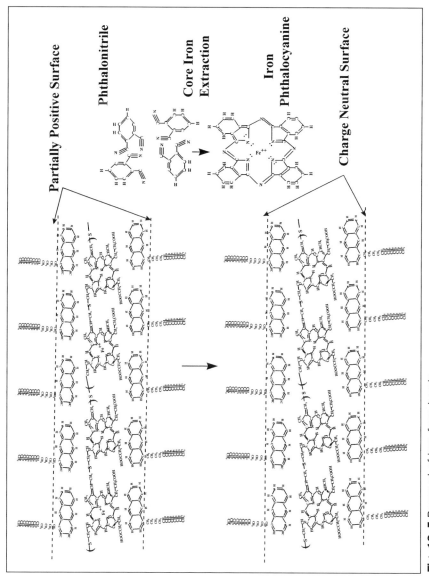

Fig. 13-5 Removal of iron from hemin

What type of catalytic reaction can be used to cleave the polysulfide links without being too costly or too drastic? Di-alkyl sulfides are susceptible to oxidation reactions, free radical reactions, and acidic reactions. However, oxidation reactions are dangerous and require considerable amounts of reactant to affect a chemical reaction. Much of the oxidizing agent would also be spent on the iron complex and other reactive compounds present in the crude. Reducing agents and conditions are also considerably more drastic than those typically encountered in production operations. This leaves the radical reaction procedure as the only realistic possibility.

Radical Reactions with Sulfur

Radical reactions with sulfur, and substituted alkyl sulfur compounds, are very common in polymer technology. Mercaptans ($C_nH_{2n+1}SH$) have long been known for their chain transfer capability and are commonly employed as molecular weight modifiers in polymerization reactions of vinyl monomers. The hydrogen of the mercaptan is readily abstracted by terminal radicals of the growing vinyl polymer chain. This abstraction process terminates the growth of the polymer with hydrogen provided by the mercaptan.

The resulting alkyl sulfide radical can then go on to initiate polymerization of a new vinyl polymer chain, form a di-sulfide by radical-radical reaction with another mercaptan radical, or abstract a proton from another proton laibile source. Sulfur and mercaptan behavior, as described above, are also consistent with the asphaltene core composition. The problem, however, remains unsolved. What type of radical reaction can take place to cleave the sulfide links without being spent on the metals present at conditions present in the field?

The radical used to attack the polysulfide links of the core asphaltene must be energetic enough to penetrate the protective sheath, cause heterolytic cleavage of the R-S-R, and produce a hydrogen-terminated R-SH. It must also be insensitive to metal radical interaction and capable of producing a radical that continuously propagates the desired cleavage (be truly catalytic).

This excludes peroxides, because they are reducible by the iron, and produce various forms explosive ethers. Azonitriles are relatively insensitive to metals, and decompose on heating to form reasonably energetic radicals, and therefore could be considered. They are very sensitive to decomposition by heat, and prolonged periods of storage at moderate temperatures. One additional possibility involves the use of an asymmetrically substituted di-alkyl sulfide. Figure 13–6 illustrates the formation (and self-propagation potential) of vinyl benzyl tertiary mercaptan.

Combination Product for Treatment of Asphaltene

At this point in the discussion we have proposed possible answers to the questions posed at the beginning of this chapter. All of these proposed answers are only suspected to be valid, since they have not to date been

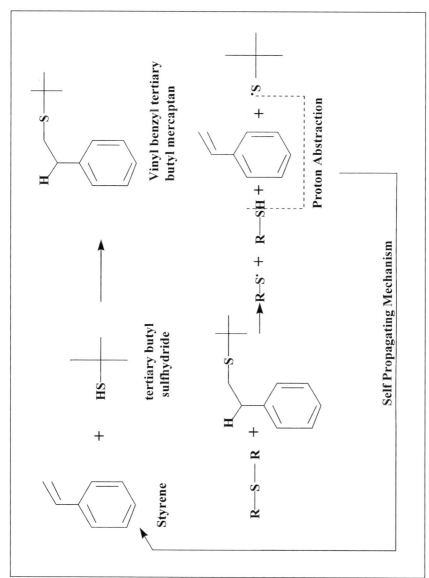

Fig. 13–6 Formation and self-propagation of vinyl benzyl tertiary mercaptan

applied in real asphaltene systems. The validity of our proposed structure for asphaltene aggregates, the reactivity of the proposed additives, and the end products of these reactions serve only as a model for the methods used to develop chemistries for the treatment of complex surfaces.

However, if the approach taken proves to have merit, it can be extended to include other possible approaches to solutions. Thus, from the above discussion a system of products including analogous chemistries to the polyvinyl pyrrolidone, phthalonitrile, and vinyl benzyl tertiary butyl mercaptan could be combined and used to treat problems of asphaltene deposition.

Summary

The preceding chapter discussed some of the methods used for the characterization of asphaltenes. The main emphasis in these test methodologies is the determination of the physical behavior of these systems under different conditions of pressure, temperature, and shear. Additional methods were suggested for the determination of molecular weight, microwave techniques, and complex formation analysis. This chapter also emphasized the relatively poor understanding of the chemical nature of asphaltenes that has resulted in the very limited methods of analysis.

The preceding chapter was also intended to illustrate the thought process involved in developing chemistries aimed at the solution of a specific problem. The starting point for this process is the formulation of appropriate questions and a rational approach to answering the questions posed. Another aspect of this approach involves the validity of the proposed structure of asphaltenes. The entire scheme could fall like a house of cards if the structure is vastly different than the one proposed. Nevertheless, research involves the taking of risks, but early preparation and thoughtful approaches reduce the risks involved.

References

Barrow, Gordon M. *Physical Chemistry*. 2d. ed. New York: McGraw-Hill Book Co., 1966.

Lehninger, L. Albert. *Biochemistry: The Molecular Basis of Cell Structure and Function*. New York: Worth Publishers, Inc., 1970.

Rodriguez, Ferdinand. *Principles of Polymer Systems*. New York: McGraw-Hill Book Co., 1970.

Seymour, Raymond B. *Introduction to Polymer Chemistry*. New York: McGraw-Hill Book Co., 1971.

14
Physical Properties of Treating Chemicals

Some Physical and Chemical Testing

One of the most difficult problems to solve prior to the field application of any treating chemical deals with the formulation of chemistries that effectively treat problems within the constraints of the system. Satisfactory physical characteristics of the treating chemical are included as a part of the constraints the system demands. If a chemical freezes at 50° F, its application in areas where the average ambient temperature is below 30° F is ill-advised, since it would be frozen most of the time. Most chemicals that are added to wells, transfer lines, or storage facilities are pumped, and frozen materials will not lend themselves to pumping. Therefore, blending of the active ingredient is necessary to facilitates its injection by pumps.

The process of blending can be very involved and requires a great deal of expertise and physical testing methods to assure a compatible product is provided to the usage point. Many of the test procedures can be applied to the entire product line of a specialty chemical company, and specific chemicals require some special tests. In this chapter an effort will be made to list some of the methods used by specialty treating chemical companies to determine if the product/products are suitable for shipment. Generally this function is performed by a quality control group for the purpose of assuring that the product to be shipped falls within certain guidelines of chemical and physical behavior.

Chemical Tests

Hydroxyl numbers

During the manufacturing of propylene and ethylene oxide products, it is necessary to determine if the reaction has proceeded to the synthetic specifications of the product development chemist. When a parent molecule is propoxylated to form the second step product in a multiphase oxide product, it is necessary to monitor the extent to which the reaction has proceeded. Just after the manufacturing procedure has reached a theoretical charge of oxide, and the pressure has receded, a sample is taken to the laboratory for analysis.

One type of an analysis that is performed employs acetic anhydride to acylate the terminal hydroxyl units present in the molecule. This reaction is performed by the addition of a pyridine acetic anhydride mixture to a weighed amount of product. The pyridine/acetic anhydride is added such that there is an excess of acetic anhydride to the hydroxyl functional of the oxide adduct. The product/acetic anhydride pyridine mix is heated to reflux to assure the reaction is complete. Once the reaction is complete, water is added, and the acetic acid formed from the hydrolysis is titratated against a blank (a sample of refluxed acetic anhydride/pyridine minus sample) with a standard normality of sodium hydroxide.

The acetate ester of the hydroxyl function of the sample is not available for titration, and the difference between the blank and the sample is taken to represent the degree to which the epoxidation reaction has proceeded. This test is commonly referred to as hydroxyl number testing.

Nuclear magnetic resonance spectroscopy

Another method that has been employed for control of oxide additions involves the use of nuclear magnetic resonance spectroscopy (NMR). This technique is particularly valuable when parent molecules such as phenolic resins are employed as the nucleus for epoxidation reactions. The aromatic proton shifts are sufficiently removed from the methylene, methine, and methyl protons that a ratio can be determined by integrating the peaks at each absorption frequency. Figure 14–1 shows a representation of this technique.

There are limitations to this method, and one of the most significant arises when the parent molecule is the same as the subunits added through the propoxylation. A good example of this is found when di-propylene glycol is used as the parent, which is often the case when high molecular weight poly propylene oxide products are made. However, the NMR analysis proves valuable when mixed propylene/ethylene oxide reactions are performed, since the methyl protons of the propylene can be compared to the methine and methylene protons adjacent to the oxygen.

Gel permeation chromatography

Molecular weights of water in oil de-emulsifiers have been correlated to their efficiency in the resolution of extremely small (stable) emulsions, thus a measure of the molecular weight of certain products has become an important

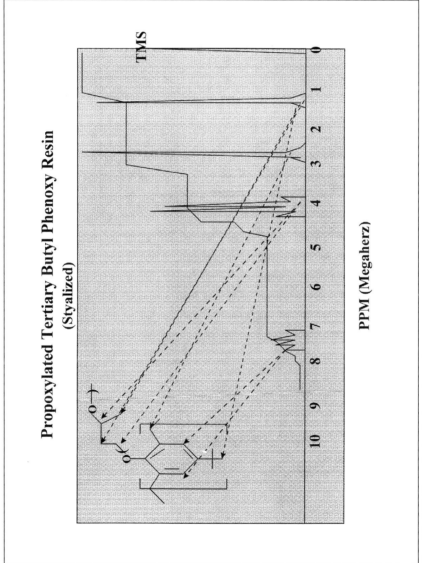

Fig. 14–1 NMR analysis of phenolic resin

part of quality control. Gel permeation chromatography (GPC) is widely used for this determination. Naturally occurring macromolecules have been fractionated for years by use of columns packed with aqueous dextran gels. This separation is based on the selective separation of the polymer molecules, which penetrate the rigid gel structure constituting the stationary phase in these columns.

Crosslinked polystyrene resins are swollen by appropriate solvents in order to utilize this technique for the separation of many synthetic polymers. By the use of appropriate liquid-solid elution columns, it is possible to characterize molecules as small as helium and as large as macromolecules with an average molecular weight of 100 million. Commercial GPC units employ the change in the index of refraction with concentration for specific polymers in dilute solution. Sigmoidal curves are obtained when the elution volume is plotted against the logarithm of molecular weight.

Physical Testing
Freeze-thaw stability

The stability of field treating chemicals under widely variable conditions of moisture level, environmental temperature, and applied pump shearing forces plays a very important role in determining whether the chemical will be suitable for sale. Because of this, physical tests are conducted that are exaggerated approximations of the conditions to which the chemicals will be subjected. Freeze-thaw testing is one such test, and it consists of exposing the chemical to cycles of temperature that are extremes of those expected in field applications. The sample is placed in an oven at 140° F and held there for a minimum of 1 day. After the sample has been exposed to 140° F for 1 day it is placed in a freezer at -40° F, held for 1 day, and the cycle is repeated at least 4 times. The sample must not separate into phases or change in appearance over this period, or it will be rejected.

Sedimentation stability

Polymer molecules remain suspended under ordinary conditions because the kinetic energy of the solvent molecules (Brownian motion) is very much greater than the sedimentation of the force of gravity. However, aggregations can form between polymer molecules and settling can occur. Thus the technique of ultracentrifugation is used to force these aggregations to occur, and the resultant increase in aggregate mass increases the likelihood of sedimentation. This technique is usually performed on capillary tubes of sample, and after a period of several minutes at high speed, the sample is visually inspected for phase separations.

Viscosity measurements

When chemical treatments are applied in field applications, they are usually applied through the use of pumps. The rheology of the product is an important factor in the ability of the pump to move the product. If the viscosity is too high, the pump will cavitate, stop, shear thin the product if it is thixotropic, or shear thicken the product if it is dilatant. Thus viscosity profiles are taken on the product under varying conditions of shear stress and temperature to determine if adverse physical behavior will result upon pumping the chemical.

Specialized testing procedures

Very often specialized tests are conducted on the products to determine their performance effectiveness in the applications for which they were designed. Water in oil emulsion breakers are routinely screened for performance on well characterized emulsion systems, which are obtained on a regular basis from nearby production areas. The testing is performed as described in chapter 5, where the crude sample is injected with a range of chemical from 10 ppm to 100 ppm, shaken vigorously for 5 minutes, placed in a water bath at 140° F, and visually observed for rate of water drop, smoothness of the interface, and completeness of treatment.

Oil in water de-emulsifier testing is less frequently performed, but can be valuable in gauging the effectiveness of these products. Synthetic oil in water emulsions can be produced by adding a sample of crude oil to a brine solution, and agitating it with a high shear device (e.g., a blender). Once the emulsion is formed, the chemical/chemicals are added in dosage levels from 1 ppm to 20 ppm and visually observed for phase separation. The clarity of the water is usually the determinant of the chemical's judged effectiveness. However, solvent extraction methods described in chapter 5 are also used.

Summary

The preceding chapter discussed a very limited number of tests that are routinely performed by quality control laboratories of specialty chemical companies. The intent of the chapter was to illustrate the extent to which control of the chemistry is monitored, but not as an exhaustive listing of the testing performed by these laboratories. Obviously, procedures for the analysis of amines, quaternary ammonium salts, phenolic resin molecular weight, aldehyde content, acid content, base content, conductivity, and a host of other chemical and physical properties are performed.

Specific mention was made of the hydroxyl, NMR, and GPC methods, because of the high proportion of products the epoxide chemistry represents to the specialty chemical companies. Representative physical test also were mentioned to emphasize the precautions necessary for the provision of chemistries with high resistance to physical changes to the end user.

References

Barrow, Gordon M. *Physical Chemistry*. 2d. ed. New York: McGraw-Hill BookCo., 1966.
Lehninger, L. Albert. *Biochemistry: The Molecular Basis of Cell Structure and Function*. New York: Worth Publishers, Inc., 1970.

Rodriguez, Ferdinand. *Principles of Polymer Systems.* New York: McGraw-Hill Book Co., 1970.

Seymour, Raymond B. *Introduction to Polymer Chemistry.* New York: McGraw-Hill Book Co., 1971.

Appendix A
Mixing Forces

The state of equilibrium conditions in the reservoir does not preclude the existence of emulsions that may have formed in the process of percolation through the tortuous pore spaces within the formation. The resulting emulsions are limited in size to aggregates that possess sufficient surface tension to remain intact under prevailing conditions. The existence of stable emulsions demands that their size be limited to one allowing aggregate velocities sufficient to balance gravitational settling forces, but insufficient to disrupt the boundary forces maintaining emulsion structure.

$$F_a > m_a(dv_a/dt) => m_a g$$

where

F_a	represents the boundary forces
m_a	is the aggregate (emulsion) mass
dt	is the time change
v_a	is the aggregate velocity
g	is the acceleration of gravity

Thus, if stability is to exist $(dv_a/dt) => g$, or the magnitude of aggregate velocity changes with time must be equal to or greater than gravitational acceleration. Further, the collision forces between emulsions and other surfaces must be less than the interface stabilization forces.

$$(F_a/A_e) > (m_a/A_e)(dv_a/dt) \Rightarrow (m_a/A_e)g$$

where

A_e =   emulsion surface area.

The term (F_a/A_e) defines the surface tension of the emulsion in units of dynes/cm. Assuming a sphere shape for the emulsion (representing minimum surface exposure), and stipulating equality between the first and third terms, a rearrangement of the relation allows expression for the maximum coefficient of emulsion size.

$$[3F_a/(4\pi m_a g)]^{1/3} = \lambda_r$$

where

λ_r is the maximum allowable coefficient of radius for a stable emulsion (radial coefficient)

When the forces balance, the left side of the expression reduces to $(3/(4\pi))^{1/3} \approx .62$. This can be taken to mean that the force of gravity acting on the emulsion cannot exceed 62% of the stabilization force of the emulsion. Then integrating and multiplying by time gives the maximum stable emulsion radius.

$$t\int dv_a < (4\pi g/3)(.62)\int t\, dt = r$$

Finally from the kinetic theory we can derive an expression for an average velocity, v_a, as follows:

$$v_a \sim (3kT/2m_a)^{1/2}$$

$$\int dv_a = (3k/2\, m_a)^{1/2} \int T^{1/2}\, dT$$

where

k is Boltzmann's constant
T is absolute temperature

Given this relationship, we would expect the radius to be fairly approximated by the following expression:

$$r \approx t(3k/2\, m_a)^{1/2} \int T^{1/2} dT \approx (4\pi g/3)(.62)\int t\, dt$$

This expression clearly indicates that the radius of the emulsion varies directly as a function of time at a temperature, and inversely as the square root of the emulsion.

Appendix B

The emulsion radius can be approximately determined by the following equation:

$$r \approx 4/3 \; \pi g \gamma \int t \, dt = 4/3 \; g\pi \; (.62) \int t \, dt$$

If the emulsion is considered as the source of main resistance to current flow, then an equivalent circuit can be constructed as in Figure B–1.

From this equivalent circuit, an expression for the inductive time constant is shown to be the following:

$$\tau_L = L/R$$

$$I = V_2/R \; (1 - e^{t/\tau})$$

If we put $t = \tau_L = L/R$ in the second expression above, this expression reduces to the following:

$$I = V_2/R(1 - e^1) = 0.63 \; V_2/R$$

Thus the time constant τ_L is the time at which the current in the circuit will reach a value within $1/e$ (about 37%) of its final equilibrium value. Note, however, that $I = 0.63 \; V_2/R \approx 0.62 \; V_2/R = \gamma \; V_2/R$, from the expression for the stable emulsion radius. Thus we could substitute RI/V_2 for γ in the expression for the radius of a stable emulsion that results in the following expression:

$$r \approx 4/3 g\pi \; RI \; / \; V_2 \int t \, dt$$

$$r \approx 4/3g\pi \; RI \; / \; V_2 \int 1/v \; dv$$

where

v \qquad is equal to $(1/2\sqrt{(1/LC)}) = f$ (see Figure B–1).

This can be interpreted to mean that the radius of the emulsion will change as a function of the resistance times current, divided by the applied voltage x force of gravity x the inverse square of the frequency. Or, more simply, the emulsion will remain stable up to a ratio of $(RI \; / \; V_2) = .62$, which is in agreement with the thickness given by the theory of the Helmholtz double layer. In Introduction to Colloid Chemistry, Karol Mysels states, "This thickness is generally taken as the distance over which the potential drops to a certain fraction of its value, specifically to $1/e = .37$ of the value." As a result of this time average distortion being imposed on emulsions by a high voltage alternating current source, devices like electrical desalters can be used to resolve emulsions. Finally, using Stokes' law expression and the first expression for the radius above, we can derive an expression for the settling velocity of an emulsion in an electric field.

$$V = \{(4\pi^2 \; g^3/(27\eta V_2^2)\} \; (RI(\Delta t))^2(d_1 - d_2)$$

$$V = \{(4\pi^2 \; g^3/(27Ae^{\Delta Evisc/RT} \; V_2^2)\} \; (RI(\Delta t))^2(d_1 - d_2)$$

This last expression for settling velocity is offered as an item for those interested in further study.

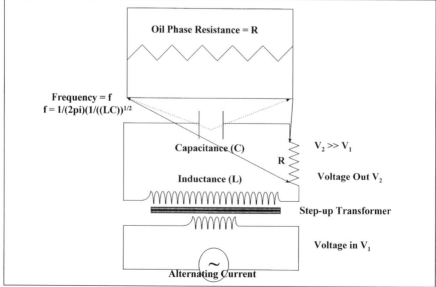

B–1 Circuit with emulsion as the main resistance to current flow

Appendix C

Molecular combinations not only require adequate momenta to overcome the surface energy barrier, they also require a minimum residence time on the opposite side of that barrier to establish cohesive forces. Molecules possessing too much momenta may pass through this energy barrier, retain an excess of inertia, and rebound through the barrier in the opposite direction. Molecules possessing too little momenta will never penetrate the energy barrier. Only molecules possessing momenta great enough to pass the barrier and retain a minimal inertia (less than that required to exit) will be afforded the time to establish the forces of attraction.

Quantization

The development of the theme that points to the quantized nature of interactions leading to higher order structural aggregates shows their dependence on statistical phenomena. As a mater of fact, the inductive effect can only be treated through quantum mechanics, since it is concerned with the detailed motions of the electrons in the neighboring molecules. Thus, a descriptive approach to some of the quantum physics involved in these interactions is appropriate and relevant to the present discussion.

In early studies of the atom, classical physical calculations dealing with the motions of the electron were found inadequate. It turns out that the extreme velocities and diminutive size of these particles, when subjected to classical mechanical calculations, gave vastly different predicted behavior than that actually observed. Thus, quantum physics was developed as a special case of classical mechanics. The challenge was to develop a theoretical model that could be used to describe the mechanics of the entire range of sizes and speed from the very small (the elementary particles, electrons, protons, neutrons etc.) to the very large (planets, stars, and galaxies).

A key development leading to the construction of this model was the discovery by Max Planck that energy is emitted from the atom in discrete packets called quanta. The adsorption or emission of these discrete packets of energy as photons was recognized, by Niels Bohr, as a means of describing why electrons orbit the atom at certain allowed radii. But there was a fly in the ointment: these energy packets consisted of photons, and photons possess no mass, so how could a massless particle have momentum?

Albert Einstein provided the answer to this question by introducing the special theory of relativity. The hypothesis that matter and energy are different forms of the same thing, or that mass and energy differ only as a function of the square of the speed of light, suggested that a unification of quantum physics and classical physics could be achieved. Bohr recognized that a ratio of Planck's constant to the wavelength of the emitted photon could be equated to a ratio of Einstein's energy to the speed of light, and that this was the momentum of the photon $(h\upsilon/\lambda=E/C)$.

Einstein's theory, however, also developed several other relationships that have far-reaching consequences on the development of a comprehensive description of the behavior of physical systems. The inclusion of time as a dimension is probably the most important. Time, as a dimension, demands that a second condition of existence be added to the description of position in space. It also changes the expression that "no two objects may occupy the same three spacial coordinates at the same time" to "no two objects can occupy the same space-time coordinate."

Invoking the special theory of relativity to describe the behavior of physical objects yields some seemingly paradoxical results. It seems as physical objects approach the speed of light their masses approach infinity, and their sizes (dimensions) are contracted. This means that time, being included as a dimension, also undergoes a real contraction. The seeming paradox arising from this view can be illustrated by an example:

> One brother of a pair of twins is given the opportunity to travel to a nearby star at near the speed of light. He travels to the nearby star, visits a planet in its system, and returns at nearly the speed of light. During his flight he aged 10 years, but when he arrived home his brother, his brother's son, and a thousand generations had lived and died.

This example seems to contradict all our conceptions of time, but this is more a result of our familiarity with exceedingly small velocities in relation to that of light. The verification of the validity of this contraction is proved when a short-lived particle created by collision of cosmic radiation in the upper atmosphere manages to reach observers on the surface of the earth. The life span of a muon at rest has been measured and determined to be 10^{-6} seconds. If a muon is created in the upper atmosphere and travels some

300 miles to a ground observer, even at the speed of light it would take $300/186,000$ or 1.61×10^3 seconds. This time of travel would then be equivalent to ~1600 muon life times. Thus the muon's arrival and its observations at the Earth's surface can best be explained by the phenomenon of time dilation. Each inertial system comprises a unique time, and as x, y, and z positions change, so does t.

Synchronosity and Temperature

From the discussion above, it is seen that the distinct or individually unique characteristics of a set of space-time coordinates are inherent to inertial systems. A less obvious, but perhaps more profound, consequence of this condition is that space-time coordinates have no meaning unless occupied by mass/energy. Or it might be said that any description of physical phenomena exclusive of the time dimension is fruitless.

Furthermore, the concept of simultaneity or synchronous time is at best an approximation, and this approximation of the time coordinate of one inertial frame of reference to another is close only when the relative velocities of each to the other are small. Thus, Newtonian or classical mechanics fits well with observation at modest velocities, but relativistic/quantum mechanics must be invoked to explain the observations of very high-speed systems.

So what does all this have to do with the forces that allow emulsion formation, or crystal formation? Remember our discussion of London forces of induction. According to Gordon Barrow, "All molecules, including those without a permanent dipole, attract each other. The liquid state, for example, is exhibited by all compounds and even the noble gasses." Since these forces arise from the hybridization of electron orbitals, and the speeds of electrons in these orbitals are time-dependent phenomena, it is necessary to invoke quantum mechanics to describe the interactions.

Thus, the blending of individual molecular orbitals requires the repositioning of electrons to allowed energy levels, and an accompanying promotion or demotion of electron energies. The promotion of an electron to a higher energy level requires the adsorption of a photon, and conversely, the demotion of an electron to a lower level requires the emission of a photon, to preserve momentum. The law of conservation of momenta must be satisfied, and this conservation is achieved by the emission or absorption of a photon.

However, if photons are massless, how can they add to or subtract from the momenta of the system? This is why the Bohr relation is crucial to understanding how momentum is conserved. Further, the processes involved in the hybridization of orbitals take place under conditions of time spans and velocities that are extremely rapid. The magnitude of the velocities involved are in excess of 2 million m/sec, and the duration of these processes is (atomic radii /electron velocity) = $10^{-8}/(2 \times 10^6 \text{ m/sec}) \sim 10^{-15}$ sec.

Thermodynamics represents a highly organized and well-substantiated description of the macro behavior of physical systems. Key to this discipline is

the concept of temperature and its influence on the systems under study. Thermodynamicists have defined temperature (according to Noller):

for a two level system in which the energy separation of the two levels is E

$$T=(E/k)/[log(N_{off}/N_{on})]$$

where

N_{off} is the number of particles in the lower energy level
N_{on} is the number in the upper

Boltzmanns' constant "k" is a statistical term that expresses the average energy possessed by a particle at a given temperature. Considering this expression, and Maxwells' speed distribution law, the relationship between the population of energy levels by mass/energy can be understood in terms of inertial frames. So the ratio of E/k represents a state of existence or, more precisely, the ratio of observed space-time coordinates to a calculated average. This state of existence then can be conceived of as the summation of space-time displacements of the individual inertial systems from the observed. Combining this concept with that of the complex time-temperature plane described above helps to illuminate the dual characteristic of mass/energy.

Duality

Photons are said to have no mass, but considering the teachings of the theory of relativity, would the statement that photons *appear* to have no mass, since they travel at the speed of light, be more appropriate? If the momentum *(mv)* of a material object approaches the infinite, as its time dimension as well as its x, y, z dimensions are contracted infinitesimally when it approaches the speed of light, then would not $mv/(v(=c))$ imply (∞/c), which then would imply that mass = ∞.

If this infinitely massive object has its energy extended infinitely in all directions from its inertial origin, then wherever it is detected it will appear as • mass/• distance ——implying how its energy is distributed in space. Thus, it is possible to see the reason for the equivalence of mass and energy $(\Delta mc^2 = \Delta E)$, and that it is the perception of this equivalence that brings about the errors in describing its manifestations. It is pertinent at this point to state pedagogically, all matter radiates mass/energy, since it is mass/energy, and the wavelength at which it radiates is determined by its inertial space-time coordinates. Thus, the duality of physical matter is the result of our perception, not its actual existence.

Given that temperature is a manifestation of momenta, it should now be clear that the space-time coordinates of these inertial systems (momenta possessing systems) represent both spacial and temporal displacements. The question is then, are these temporal and spacial displacements separable?

And if they are, can the complex time-temperature surface be invoked to elaborate the nature of their separation? And further, can elaboration of the nature of this dimensional separation produce insight into our perception of the dual nature of matter? If time and temperature are defined as an imaginary surface, and "...we replace the time (which is represented by a *real* number, such as 5 or 320) by an imaginary number (a number of the form *ib*, such as 5i or 320i)...certain equations in ordinary dynamics are turned into *thermo*dynamics expressions..." (according to Layzer). Further, Einstein expressed the true geometric surface of space as follows:

$$X^2 + Y^2 + Z^2 + (-ict)^2 = S$$

where

i is equal to $(-1)^2$
c is the velocity of light

Combining these two statements we can then substitute temperature (T) for (it) in the above expression and derive the following:

$$X^2 + Y^2 + Z^2 + (-cT)^2 = S$$

And if this is true, then either the dimensions used to describe time or those used to describe temperature must be in error. Is this apparent error an artifact of our perception or a real physical anomaly? It should be noted that in Einstein's derivation for the time-space, surface applies to all space, and more appropriately all surfaces. Thus, it must be concluded that whenever any surface is described, it should be recognized that it is best represented by four coordinates, not three, and that when aggregation of substructures are discussed the coordinate of time must be part of the discussion.

Often the expression of the ratio of total energy to kinetic energy is used to represent the ratio of radiant to kinetic energy.

$$(hu/nkT) = (E_{rad}/K.E.)$$

where

h is joule-seconds
v is 1/seconds
n is number particles
k is joule/($^\circ$ K-number particles)
T is $^\circ$ K
E_{rad} is joules
$K.E.$ is joules

However, if just the ratio of (h/nk) is dimensionally analyzed, the units end up as (°K-sec). But if $iT = t$, then the ratio would become (sec^2 or temp.^{2}).

From the above discussion, it should be clear that the description of many physical systems tends to become blurred by our tendency to misperceive the true nature of surfaces (dimensions). This misperception results from our need to extrapolate from our observational experience of the visible world and ascribe these observed characteristics to the behavior of less easily observed physical phenomena.

References

Barrow, Gordon M., *Physical Chemistry*. 2nd. ed. New York: McGraw-Hill Book Co., 1966.

Layzer, David, *Constructing the Universe*. New York: Scientific American Library, 1984.

Noller, Carl R. *Textbook of Organic Chemistry*. 3rd ed. London: W. B. Saunders Co., 1966.

Index

A